Contemporary's

Number Power

a real world approach to math

Kenneth Tamarkin

McGraw Hill **Wright Group**

Wright Group

ISBN: 0-8092-2378-3

Send all inquiries to:
Wright Group/McGraw-Hill
130 E. Randolph, Suite 400
Chicago, IL 60601

Printed in the United States of America.

7 8 9 10 11 CUS CUS 08 07 06

The McGraw-Hill Companies

Table of Contents

Multiplication and Division Word Problems: Whole Numbers

Multiplication and Division Word Problems: Decimals and Fractions

Using Proportions

Strategies with Mixed Word Problems

Percent Word Problems

Combination Word Problems

Word Problems Posttest A

Word Problems Posttest B

USING NUMBER POWER

To the Student

Welcome to Word Problems:

This Number Power workbook is designed to help you understand how to solve word problems found in the workplace and in your everyday experiences. The first section of the book, Building Number Power, provides step-by-step instruction and practice in reading and solving word problems. The last section of the book, Using Number Power, provides further practice in solving word problems in real-life situations.

Throughout this book, you will be encouraged to use mental math. You will see word problems with small or easy to compute numbers. Once you figure out which math operation you should use, you might be able to do the math in your head.

Also scattered through the book are tips on estimation and using a calculator. The following icons will alert you to the problems where using these skills will be especially helpful.

 calculation icon

 estimation icon

 mental math icon

To get the most from your work, do each problem carefully. Check each answer to make sure you are working accurately. An answer key is provided at the back of the book. Inside the back cover is a chart to help you keep track of your score on each exercise.

Also included in the back of the book are quick reference pages for using a calculator, estimation, mental math, and formulas and measurements. Refer to these pages for a quick review and helpful explanation.

Word Problem Pretest

This test will tell you which sections of *Number Power 6* you need to concentrate on. Do every problem that you can. Round decimals to the nearest cent or the nearest hundredth. Correct answers are listed by page number at the back of the book. After you check your answers, the chart at the end of the test will guide you to the pages of the book where you need work.

1. 1,600 pounds of steel are used to make a Chevrolet. The automobile plant produced 840 Chevrolets in one day. How many pounds of steel were needed that day to make the cars?

   a. 2,440 pounds
   b. 1,344,000 pounds
   c. 760 pounds
   d. 244,000 pounds
   e. none of the above

2. In one year, Melinda grew $2\frac{1}{4}$ inches to $48\frac{3}{8}$ inches. What was her height at the beginning of the year?

   a. $50\frac{5}{8}$ inches
   b. $46\frac{1}{8}$ inches
   c. $43\frac{1}{2}$ inches
   d. $21\frac{1}{2}$ inches
   e. $108\frac{27}{32}$ inches

3. Cynthia took 19 girls roller blading. If it cost $0.75 for each of the children to get in and $0.50 for each of them to rent roller blades, how much money did Cynthia have to collect?

   a. $20.25
   b. $23.75
   c. $17.75
   d. $1.25
   e. $4.75

4. During the big spring sale, Jean bought a coat for $79.50, which was 75% of the original price. What was the original price of the coat?

   a. $106.00
   b. $59.63
   c. $154.50
   d. $94.34
   e. not enough information given

5. Diana makes lemonade from the powdered concentrate by combining 5 tablespoons of concentrate with 2 cups of water. The directions say you should use 24 cups of water for the entire container of concentrate. How many tablespoons of concentrate are in the container?

    a. 240 tablespoons
    b. 130 tablespoons
    c. 110 tablespoons
    d. 60 tablespoons
    e. 31 tablespoons

6. Carol was told that she would have to pay $684 interest on a $3,600 loan. What interest rate would she have to pay?

    a. $2,912
    b. $4,284
    c. 19%
    d. 5.3%
    e. 81%

7. Jack bought a turkey for $10.34 and a chicken for $5.17. How much did he spend on the meat?

    a. $2.00
    b. $5.17
    c. $15.51
    d. $53.46
    e. $20.00

8. The New Software Company received a shipment of 200,000 foam pellets to be used in packing boxes. If New Software uses on the average 400 pellets for each box, how many boxes can be packed using the shipment of pellets?

    a. 2,000 boxes
    b. 800 boxes
    c. 199,600 boxes
    d. 80,000,000 boxes
    e. 500 boxes

9. Oranges cost $2.40 a dozen. Winsome bought the fruit pictured here. How much money did she spend on the oranges?

    a. $7.20
    b. $2.44
    c. $2.36
    d. $0.80
    e. $0.60

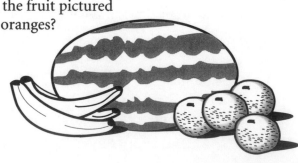

10. A roast weighing 3.15 pounds is cut into 24 slices. On the average, how much does each slice weigh?

    a. 27.15 pounds
    b. 20.85 pounds
    c. 75.60 pounds
    d. 0.13 pounds
    e. 1.31 pounds

11. Barbara needed 180 inches of masking tape to mask a window for painting. How many rolls of masking tape does she need to mask 12 of these identical windows?

    a. 18 feet
    b. 15 rolls
    c. 192 inches
    d. 168 inches
    e. not enough information given

12. A factory produces $\frac{7}{8}$-ton steel girders. How much steel does the factory need to produce 600 of the girders?

    a. 525 tons
    b. 52.5 tons
    c. 686 tons
    d. 68.6 tons
    e. none of the above

13. During the sale, Naisuon bought a three-piece wool suit that was reduced by $98 to $190. What was the original price of the suit?

    a. $92
    b. $288
    c. $276
    d. $96
    e. $291

14. Out of 1,400 people polled, 68% were in favor of a nuclear arms freeze, 25% were against it, and the rest were undecided. How many people were undecided?

    a. 93 people
    b. 350 people
    c. 952 people
    d. 98 people
    e. 1,307 people

15. A $1\frac{1}{4}$-pound lobster costs $7.80. How much does it cost per pound?

    a. $9.75
    b. $6.24
    c. $6.55
    d. $9.05
    e. $1.56

16. At the gas station, Verna tried to fill up her 18-gallon gas tank. When the tank was filled, the gasoline pump looked like the picture at right. How much gas was in the tank before Verna started pumping the gas?

    a. 0.71 gallon
    b. 71 gallons
    c. 5.22 gallons
    d. 6.78 gallons
    e. 30.78 gallons

    GAS
    | PRICE PER GALLON | 1.29 |
    | TOTAL GALLONS | 12.78 |
    | TOTAL PRICE | 16.60 |

17. During the Washington's birthday clearance sale, Gayle bought a $96 coat that was reduced by $\frac{1}{3}$. What was the sale price of the coat?

    a. $32
    b. $64
    c. $288
    d. $93
    e. none of the above

18. Shorie's rent has been increased $65 a month to $780 a month. What had she been paying?

    a. $715
    b. $845
    c. $72
    d. $9,360
    e. $50,700

19. Lillian's allergy pills come in the bottle pictured at the right. She takes four tablets a day. How many tablets did she have left after taking the tablets for 30 days?

    a. 130 tablets
    b. 216 tablets
    c. 120 tablets
    d. 370 tablets
    e. not enough information given

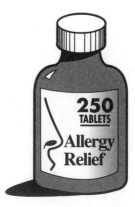

250 TABLETS
Allergy Relief

20. A cereal manufacturer puts 2 ounces of sugar in every box of cereal. How many pounds of sugar are needed for 1,000 boxes?

    a. 50 pounds
    b. 20 pounds
    c. 125 pounds
    d. 200 pounds
    e. 625 pounds

21. For the survey to be considered valid, 15% of the 6,000 questionnaires had to be returned. At least how many questionnaires had to be returned?

    a. 900 questionnaires
    b. 400 questionnaires
    c. 40,000 questionnaires
    d. 5,985 questionnaires
    e. not enough information given

22. An oil truck carried 9,008 gallons of oil. After making seven deliveries averaging 364 gallons each, how much oil was left in the truck?

    a. 174 gallons
    b. 9,379 gallons
    c. 8,644 gallons
    d. 6,460 gallons
    e. 8,637 gallons

23. In one week, Speculation Company's stock dropped in value $1\frac{7}{8}$ dollars to $8\frac{3}{4}$ dollars a share. What was the value of the stock at the beginning of the week?

    a. $2\frac{4}{5}$ dollars
    b. $16\frac{13}{32}$ dollars
    c. $6\frac{7}{8}$ dollars
    d. $10\frac{5}{8}$ dollars
    e. $9\frac{5}{6}$ dollars

24. How much would a 1.62-pound package of lamb shoulder chops cost at $2.43 a pound?

    a. $1.50
    b. $4.05
    c. $0.81
    d. $15.00
    e. $3.94

25. Money available for financial aid at Santa Carla Community College has dropped $462,000 from last year's $1,126,200. The college decided to divide the aid equally among 820 students who needed the money. How much did each student get in financial aid?

    a. $563.41
    b. $810.00
    c. $1,373.41
    d. $1,936.82
    e. none of the above

Questions 26–28 are based on the following information:

The Bargain Basement marks down clothes depending on how long they have been displayed. The day an item is put on the racks or in a bin, a date ticket and a price tag are attached to it. The chart shows the amount of discount if the date ticket is at least the listed number of days old.

Days	Markdown
10 days	10%
20 days	25%
30 days	40%
40 days	75%

26. On May 30, Sharon went shopping at the Bargain Basement. She found one blouse she wanted dated April 14 with a price tag of $18. How much did she have to pay for the blouse?

    a. 46 days
    b. $13.40
    c. $4.50
    d. $22.50
    e. $18.00

27. On July 20 at the Bargain Basement, Belquis selected five bathing suits. The red bathing suit was dated June 28 and had a $40 price tag. The floral print suit was dated July 8 and had a price tag of $30. The violet bathing suit was dated June 18 and had a price of $45. The striped bathing suit was dated July 15 and had a price of $28. The black bathing suit was dated June 2 and had a price of $60. Which bathing suit was the least expensive?

    a. the red bathing suit
    b. the floral print bathing suit
    c. the violet bathing suit
    d. the striped bathing suit
    e. the black bathing suit

28. On December 3 at the Bargain Basement in New Hampshire, Marcia bought a sweater dated November 28. She paid with a $50 bill. Since New Hampshire does not have a sales tax, she paid no tax. How much change did Marcia receive?

   a. $11.00
   b. $89.00
   c. $19.50
   d. $14.90
   e. not enough information given

29. Last month, Francisco used 445 kWh of electricity in his home. On his bill, there was a customer charge of $5.81, a delivery service charge of $0.047 per kWh, and a supplier service charge of $0.032 per kWh. Which expression could be used to calculate his total bill?

   a. 445 kWh ($5.81 + $0.047 + $0.032)
   b. $5.81 + $0.047 + $0.032
   c. $5.81 + 445 kWh ($0.047 + $0.032)
   d. 445 kWh + $5.81 ($0.047 + $0.032)
   e. 445 kWh ($0.047 + $0.032)

30. Brunilda was trying to decide between buying a digital camera and a single lens reflex (SLR) camera. The digital camera cost $499 while the SLR camera cost $249. Film for the SLR camera costs $4 for a roll of 36 pictures, plus $16 for developing. Digital pictures can be stored on a Zip disk, which costs $10 and holds 720 pictures. How many pictures would Brunilda have to take for the cost of each camera plus the pictures to be the same?

   a. 13 pictures
   b. 250 pictures
   c. 260 pictures
   d. 468 pictures
   e. 756 pictures

Word Problem Pretest Chart

If you miss more than one problem in any section of this test, you should complete the lessons on the practice pages indicated on this chart. If you miss no problems in a section of this test, you may not need further study in this chapter. However, to master solving word problems, we recommend that you work through the entire book. As you do, focus on the skills covered in each chapter.

PROBLEM NUMBERS	SKILL AREA	PRACTICE PAGES
13, 18	add or subtract whole numbers	18–39
1, 8	multiply or divide whole numbers	61–76
2, 23	add or subtract fractions	52–60
12, 15	multiply or divide fractions	80–88
7, 16	add or subtract decimals	40–51, 56–60
10, 24	multiply or divide decimals	77–79, 87–88
4, 6, 21	percents	118–136
9, 20	conversion	100–101
11, 28	not enough information	110–114
3, 5, 14, 17, 19, 22, 25, 26, 27, 29, 30	multistep word problems	137–161

Building
Number
Power

INTRODUCTION TO WORD PROBLEMS

Steps in Solving Word Problems

A **word problem** is a sentence or group of sentences that tells a story, contains numbers, and asks the reader to find another number.

This is an example of a word problem:

Last week Paula earned $194. The week before, she earned $288. What was the total amount of money she earned?

In this book, you will use five steps to solve word problems. It is important to follow these steps to organize your thinking. They will help you figure out what may seem to be a difficult puzzle. In all cases, read the problem carefully, more than once if necessary. Then follow these steps.

STEP 1 Decide what the *question* is asking you to find.

STEP 2 Then, decide what *information* is *necessary* in order to solve the problem.

STEP 3 Next, decide what *arithmetic operation* to use.

STEP 4 Work out the problem and find the solution. Check your arithmetic.

STEP 5 Finally, *reread the question* to make sure that your answer *is sensible*.

Many people can do some word problems in their heads. This is known as **math intuition** and works well with small whole numbers. Math intuition often breaks down with larger numbers, decimals, and especially fractions. Additionally, word problems of two or more steps can be even more difficult.

You should practice the five-step approach even with problems that you could solve in your head. Then you will have something to fall back on when intuition is not enough.

Step 1: The Question

After reading a word problem, the first step in solving it is to decide what is being asked for. You must find the question.

The following word problem consists of only one sentence. This sentence asks a question and contains the information that is needed to solve the problem.

EXAMPLE 1 How much did Mel spend on dinner when the food cost $20 and the tax was $1?

The question asks, "How much did Mel spend on dinner?"

The next word problem contains two sentences. One sentence asks the question, and the other sentence gives the information that is necessary to solve the problem.

EXAMPLE 2 Mary got $167 a month in food stamps for 9 months. What was the total value of the stamps?

The question asks, "What was the total value of the stamps?"

Example 3 also contains two sentences. Notice that *both* sentences contain information that is necessary to solve the problem.

EXAMPLE 3 The Little Sweetheart tea set normally costs $8.95. How much did Alice save by buying the tea set for her daughter at an after-Christmas sale for $5.49?

The question asks, "How much did Alice save?"

Sometimes the question does not have a question mark.

EXAMPLE 4 Fredi has $27 in her checking account. She wrote checks for $15 and $20. Find how much money she needs to deposit in order to cover the checks.

The question asks, "Find how much money she needs to deposit in order to cover the checks."

Underline the question in each of the following word problems. DO NOT SOLVE!

1. Last winter it snowed 5 inches in December, 17 inches in January, 13 inches in February, and 2 inches in March. How much snow fell during the entire winter?

2. To cook the chicken, first brown it for 10 minutes. Then lower the temperature and let it simmer for 20 more minutes. What is the total cooking time?

3. Find the cost of parking at the meter for 3 hours if it costs 25 cents an hour to park.

4. How many years did Joe serve in prison if his sentence of five years was reduced by three for good behavior?

5. Jenny loves to plant flowers. She has $30 to spend on flower plant flats. Find the number of flats she can buy if they cost $4.98 each.

6. The recycling plant pays $22 a ton for recycled newspaper. How much did the city of Eugene receive when it delivered 174 tons of newspaper to the recycling plant?

In each of the following word problems, the question is missing. Write a possible question for each word problem.

7. A factory needs $2\frac{1}{3}$ yards of material to make a coat. The material comes in 60-yard rolls.

8. A tablet of Extra-Strength Tylenol contains 500 milligrams of acetaminophen. The normal adult dose is 2 tablets every 4 to 6 hours, not to exceed 8 tablets in 24 hours.

9. Computer Connection slashed the price of its best selling digital camera from $799 to $549.

10. The Arctictec textile factory, which operates round the clock, can produce 42,000 yards of Arctictec fabric in a 24-hour day.

11. In a normal week, Great Deal Used Cars sells 38 cars. During the big Presidents' Week sale, Great Deal sold 114 cars.

12. The snack bar menu, shown at the right, was posted on the wall.

Super Pretzels	$1.50
Chips	$.65
Nachos	$2.25
Soda (12 oz can)	$.80
Hot Dog	$2.00
Hamburger	$2.50
Pizza Slice	$1.75
Milk	$.75

Step 2: Selecting the Necessary Information

After finding the question, the next step in solving a word problem is selecting the **necessary information.** The necessary information consists of the **numbers** and the **labels** (words or symbols) that go with the numbers. The necessary information includes *only* the numbers and labels that you need to solve the problem.

The labels make the numbers in word problems concrete. For example, the necessary information in Example 1 below is not just the number *5*, but includes *5 apples.* Paying close attention to labels will help you learn many of the methods shown in this book and will help you avoid common mistakes with word problems.

After each of the following examples, the necessary information is listed.

EXAMPLE 1   Doreen bought 5 apples last week and 6 apples this week. How many apples did she buy altogether?

The necessary information is *5 apples* and *6 apples*. Both numbers are followed by the label word *apples*.

EXAMPLE 2   A shirt costs $9.99. What is the cost of 5 shirts?

The necessary information is *$9.99* and *5 shirts*. The labels are the dollar sign (*$*) and the word *shirts*.

EXAMPLE 3   Four ounces of detergent are needed to clean a load of laundry. How many more loads of laundry can you clean if you buy the large bottle of detergent rather than the small bottle shown at right?

The necessary information is *4 ounces, 64 ounces,* and *96 ounces*. The numbers are followed by the label word *ounces*. Note that you get some of the necessary information from the pictures.

64 ounces   96 ounces

In each word problem, find the necessary information. Circle the numbers and underline the labels. Then write the label that would be a part of the answer, but DO NOT SOLVE!

1. On Friday a commuter train took 124 commuters to work and 119 commuters home. How many commuters rode the train that day?

2. There are 14 potatoes in the bag at the right. What is the average weight of each potato?

3. Unleaded gasoline costs $0.06 more per gallon than regular. Regular costs $1.47 a gallon. How much does unleaded gasoline cost?

4. To make the punch, Lona combined the bottle of ginger ale with the container of fruit juice shown at the right. How much punch did she make?

5. The radio station added $38 more to the $329 already in the superjackpot. What is the new amount of money in the superjackpot?

6. Frank bought the package of loose-leaf paper shown at the right and put 60 pages in his binder. How many pages were left in the package?

Necessary vs. Given Information

Sometimes a word problem contains numbers that aren't needed to answer the question. You must read problems carefully to choose only the necessary information.

Notice this important difference: The **given information** includes *all* of the numbers and labels in a word problem.

The **necessary information** includes *only* those numbers and labels needed to solve the problem.

EXAMPLE 1   Nelson travels to and from work with 3 friends every day. The round trip is 9 miles. If he works 5 days a week, how many miles does he commute in a week?

given information: 3 friends, 9 miles, 5 days
necessary information: 9 miles, 5 days

To figure out how many miles he commutes in a week, you do not need to know that Nelson travels with 3 friends.

EXAMPLE 2   There are 7,000 people living in Dry Gulch. Of the 3,000 people who are registered to vote, only 1,700 people participated in the last election. How many registered voters did not vote?

given information: 7,000 people, 3,000 people, 1,700 people
necessary information: 3,000 people, 1,700 people

All of the numbers have the same label—*people*. However, the total number of people in the town (7,000) is not needed.

Sometimes you will have to choose necessary information from a chart or picture containing other information as well.

EXAMPLE 3   According to the chart, how many hours did Eduardo work on Friday and Saturday?

given information: 4 hours, 2 hours, 6 hours, 8 hours
necessary information: 6 hours, 8 hours

Hours Worked	
Monday	4
Wednesday	2
Friday	6
Saturday	8

In this book, you will practice choosing information from charts and pictures.

This exercise will help you tell the difference between given and necessary information. Underline the given information. Circle the necessary information. DO NOT SOLVE!

1. Mona is 22 years old. She has a sister who is 20 years old and a boyfriend who is 23. How much older is Mona than her sister.

2. Rena receives $186 a month from welfare. She also receives $167 a month in food stamps in order to help feed her two children. How much public assistance does she receive each month?

3. Marilyn works three times as many hours as her 20-year-old sister Laura. Laura works 10 hours a week. How many hours a week does Marilyn work?

4. Suzanne has a 7-year-old car. According to the chart at the right, how much did she spend on gasoline during the first 2 months of the year?

Gasoline Expenses	
January	$43
February	$39
March	$40
April	$31

5. During the winter, the Right Foot shoe store spent $2,460 for oil heat and sold $35,800 worth of shoes. If oil costs $1.20 per gallon, how many gallons did the shoe store buy?

6. Erma, who is 45 years old, cooks dinner for the eight people in her family. Her husband, Jack, cooks breakfast in the morning for only half of the family. For how many people does Jack cook?

7. In a factory of 4,700 workers, 3,900 are skilled laborers. Of the employees, 700 people are on layoff. How many people are currently working?

8. Maritza bought the bottle of cola shown at the right for $1.49. How many 12-ounce glasses can she fill from the bottle?

ADDITION AND SUBTRACTION WORD PROBLEMS: WHOLE NUMBERS

Finding Addition Key Words

In the first chapter, you worked on finding the question and the necessary information in a word problem. The third step in solving a word problem is **deciding which arithmetic operation to use.**

You will now look at word problems that can be solved by using either addition or subtraction. In this book, you will learn five methods to decide whether to add or subtract.

1. finding the key words
2. restating the problem
3. making drawings and diagrams
4. writing number sentences
5. using algebra

You will also work with making estimates and substitutions.

All of these methods are useful in understanding and solving word problems. After learning them, you may decide to use one or more of the methods that you find most helpful.

How do you know that you must add to solve a word problem? **Key words** can be helpful. A key word is a clue that can help you decide which arithmetic operation to use.

Note: *How many, how much,* and *what* are general mathematics question words, but they are not key words. They help to identify the question but do not tell you whether to add, subtract, multiply, or divide.

The following examples contain addition key words.

EXAMPLE 1   What is the sum of 3 dollars and 2 dollars?

addition key words: sum, and

The sum is the answer to an addition problem. Therefore, when the word *sum* appears in a word problem, it is a clue that you should probably add to solve the problem.

EXAMPLE 2   The small cup contains 16 ounces of soda. The large cup contains 6 more ounces. How many ounces are in the large cup?

addition key word: more

The word *more* suggests that you should add the two amounts together.

....

**In the following exercise, circle the key words that suggest addition.
DO NOT SOLVE!**

1. Karen bought a new car for $15,640 plus $4,600 for options. How much did she spend for the car?

2. Judy bought four lemons and twelve oranges. How many pieces of fruit did she buy altogether?

3. A recipe for pumpkin pie says that an extra 2 tablespoons of sugar can be added to extra sweetness. The standard recipe is below. How many tablespoons of sugar are needed for the sweeter pie?

GRANNY'S
PUMPKIN PIE
1 tsp salt
4 tablespoons sugar
2 tsp. cinnamon

4. The price of a $6 general admission ticket to the ballpark will increase $1 next year. What will be the general admission price next year?

6. When Nancy did her kid's laundry, she found the coins pictured below. How much change did she find in all?

5. Although she has lost 15 pounds, Pam wants to lose 10 more. How much weight does she want to lose altogether?

Solving Addition Word Problems with Key Words

You have now looked at the first three steps in solving a word problem.

> **STEP 1** Finding the question
> **STEP 2** Selecting the necessary information
> **STEP 3** Deciding what arithmetic operation to use

The next step in solving a word problem is doing the arithmetic. People who have not learned to think carefully about word problems may rush into doing the arithmetic and become confused. However, the three steps before doing the arithmetic and the one step after provide a good way to organize your thinking to solve a problem. The actual arithmetic is only one of several necessary steps.

Sometimes the arithmetic can be done in your head as **mental math.** Here are some examples that can be done as mental math.

EXAMPLE 1 What is the sum of 3 dollars and 2 dollars?

> **STEP 1** *question:* What is the sum?
>
> **STEP 2** *necessary information:* 3 dollars, 2 dollars
>
> **STEP 3** *addition key words:* sum, and
>
> **STEP 4** *add:* 3 dollars + 2 dollars = **5 dollars**

$$\begin{array}{r} 3 \\ + 2 \\ \hline 5 \end{array}$$

EXAMPLE 2 The small cup contains 16 ounces of soda. The large cup contains 6 more ounces. How many ounces are in the large cup?

> **STEP 1** *question:* How many ounces are in the large cup?
>
> **STEP 2** *necessary information:* 16 ounces, 6 ounces
>
> **STEP 3** *addition key word:* more
>
> **STEP 4** *add:* 16 ounces + 6 ounces = **22 ounces**

$$\begin{array}{r} 16 \\ + 6 \\ \hline 22 \end{array}$$

Once you have completed the arithmetic, there is one last step: reread the question and make sure that the answer is sensible.

For instance, if you had subtracted in Example 2, you would have gotten an answer of 10 ounces. Would it have made sense to say that the smaller cup was 16 ounces and the larger cup was 10 ounces?

**For each problem, circle the key word or words and do the arithmetic.
Be sure to include the label as part of your answer.**

1. After 5 inches of snow fell on a base of 23 inches of snow, how many inches of snow were on the ski trail altogether?

2. According to the chart at the right, what was the new bus fare to Georgetown after the 1999 fare was increased by 20 cents?

Bus Fares 1999	
Holliston	$.85
Green Borough	$1.75
Georgetown	$1.80
Filmore	$2.60

3. What is the total weight of a 3,500-pound truck carrying a 720-pound load?

4. How big an apartment is the Dao family looking for if they want one that is 2 rooms larger than their 3-room apartment?

5. After the church raised $121,460 the first year and $89,742 the second, how much money was in its building fund?

6. According to the price chart at the right, how much does it cost to buy both a sofa and a reclining chair?

Furniture	
Sofa	$529
Love seat	$319
Reclining chair	$449
Coffee table	$199

Finding Subtraction Key Words

Each of the key words in the previous exercises helped you decide to add. Other key words may help you decide to subtract. Here is an example of a word problem using subtraction key words.

EXAMPLE   The large cup contains 16 ounces of soda. The small cup contains 6 ounces less than the large cup. How many ounces does the small cup contain?

STEP 1   *question:* How many ounces does the small cup contain?

STEP 2   *necessary information:* 16 ounces, 6 ounces

STEP 3   *subtraction key words:* less than

> **Note:** In some subtraction word problems, the key words *less* and *than* are separated by other words. This is also true for *more* and *than*.

In the following exercise, circle the key words that suggest subtraction. DO NOT SOLVE!

1. Bargain Airlines is $25 cheaper than First Class Air. First Class charges $200 for a flight from Kansas City to St. Louis. What does Bargain Airlines charge?

2. Out of 7,103 students, State College had 1,423 graduates last year. This year there were 1,251 graduates. What was the decrease in the number of graduates?

3. The small steak weighs 5 ounces less than the large steak. How much does the small steak weigh?

12 oz

$4.59 $2.99

4. The large engine has 258 horsepower. The economy engine has 92 horsepower. What is the difference in the horsepower between the two engines?

5. This year Great Rapids has 15 schools. Next year the number of schools will be reduced by 2. How many schools does the city plan to open next year?

Solving Subtraction Word Problems with Key Words

You can now complete the example in the "Finding Subtraction Key Words" lesson.

EXAMPLE 1 The large cup contains 16 ounces of soda. The small cup contains 6 ounces less than the large cup. How many ounces does the small cup contain?

STEP 1 *question:* How many ounces does the small cup contain?

STEP 2 *necessary information:* 16 ounces, 6 ounces

STEP 3 *subtraction key words:* less than

STEP 4 *subtract:* 16 ounces − 6 ounces = **10 ounces**

$$\begin{array}{r} 16 \\ -\ 6 \\ \hline 10 \end{array}$$

In Example 2, the key words *less* and *than* are separated.

EXAMPLE 2 How much less does a $13 polyester dress cost than a $24 cotton one?

STEP 1 *question:* How much less does a $13 polyester dress cost?

STEP 2 *necessary information:* $13, $24

STEP 3 *subtraction key words:* less than

STEP 4 *subtract:* $24 − $13 = **$11**

$$\begin{array}{r} 24 \\ -\ 13 \\ \hline 11 \end{array}$$

This is also an example of a common type of subtraction problem. To solve it, you must reverse the order in which the numbers appear in the problem.

In each problem below, circle the key words and do the arithmetic. Be sure to include the label as part of your answer.

1. After a tornado destroyed 36 of the 105 homes in Carson, how many homes were left?

2. How much change did Mel receive when he paid for $16 worth of gas with a $20 bill?

3. After spending $325 of the $361 in her savings account for Christmas presents, how much did Carmena have left in her account?

4. What is the difference in price between the two cars below?

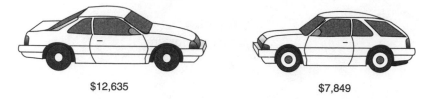

$12,635 $7,849

5. In the evening, the temperature had fallen 12 degrees from the afternoon high of 86 degrees, following a morning low of 58 degrees. What is the evening temperature?

6. Harold weighs 161 pounds, and his wife Nora weighs 104 pounds. How much more does Harold weigh than Nora?

7. Yesterday the hurricane was reported to be 420 miles offshore. Overnight it came 140 miles closer. How far from shore was it at dawn?

8. Caroline's phone bill was $121 in March and $46 in April. By how much did her phone bill decrease in April?

9. After the fire, Peg discovered that out of 460 books in her personal library, only 133 remained. How many of her books were lost in the fire?

10. Lou bought a 21-pound sirloin strip. After the butcher trimmed the fat and cut the strip into steaks, the weight of the meat was 17 pounds. How much less was the weight of the meat after the fat was trimmed?

Key Word Lists for Addition and Subtraction

Now you know that some key words may help you decide to add. Other key words may help you decide to subtract.

Here are some important key words to remember. You may want to add more words to these lists.

ADDITION KEY WORDS

sum	raise
plus	both
add	combined
and	in all
total	altogether
increase	additional
more	extra

SUBTRACTION KEY WORDS

less than	left
more than	remain
decrease	fell
difference	dropped
reduce	change
lost	
nearer }	
farther }	other *-er* comparison words

Solving Addition and Subtraction Problems with Key Words

Now that you have seen how key words are used in addition and subtraction word problems, look at the following examples to see the difference between addition word problems with key words and subtraction word problems with key words.

EXAMPLE 1 It snowed 7 inches on Monday and 5 inches on Friday. What was the total amount of snow for the week?

STEP 1 *question:* What was the total amount of snow?

STEP 2 *necessary information:* 7 inches, 5 inches

STEP 3 *addition key words:* and, total

STEP 4 *add:* 7 inches + 5 inches = **12 inches**

$$\begin{array}{r} 7 \\ + 5 \\ \hline 12 \end{array}$$

EXAMPLE 2 The city usually runs its entire fleet of 237 buses during the morning rush hour. On Thursday morning, 46 buses and 13 subway cars were out of service. How many buses were left to run during the Thursday morning rush hour?

STEP 1 *question:* How many buses were running Thursday morning?

STEP 2 *necessary information:* 237 buses, 46 buses (13 subway cars is not necessary information.)

STEP 3 *subtraction key word:* left

STEP 4 *subtract:* 237 buses − 46 buses = **191 buses**

$$\begin{array}{r} 237 \\ - 46 \\ \hline 191 \end{array}$$

In addition word problems, numbers are often being combined, and you are looking for the total. In subtraction word problems, numbers are being compared, and you are looking for the difference.

In this exercise, circle the key words. Decide whether to add or to subtract. Then solve the problem.

1. A book saleswoman sold 86 books on Monday and 53 books on Tuesday. How many books did she sell altogether?

2. After selling 15 rings on Wednesday, a jeweler sold 31 rings and 4 necklaces on Thursday. How many more rings did she sell on Thursday than on Wednesday?

3. At the town meeting, votes are recorded on the vote tally board as shown at the right. What was the total vote?

Vote Tally Board	
Yes	564
No	365

4. This year the Graphics Computer Company sold 253 units. Last year it sold 421 units. By how many units did sales decrease this year?

5. Mammoth Oil advertises that with its new brand of oil, a car can be driven 10,000 miles between oil changes. With Mammoth's old oil, a car's oil had to be changed every 3,000 miles. How much farther can you drive with Mammoth's new oil than with its old oil?

6. Last year, the Gonzales family paid $530 a month for rent. If their rent was increased by $35 a month, how much monthly rent are they now paying?

7. In April the Gonzales family paid $26 for electricity. In July their bill rose to $42. How much more did they pay in July than in April?

MORE ADDITION AND SUBTRACTION WORD PROBLEMS: WHOLE NUMBERS

Key Words Can Be Misleading

So far you have seen one approach to solving word problems.

STEP 1 Find the key word.

STEP 2 Decide whether the key word suggests addition or subtraction.

STEP 3 Do the arithmetic the key word directs you to do.

This approach can work in many situations.

But Be Careful!

Sometimes the same key word that helped you decide to add in one word problem can appear in a problem that requires subtraction.

The next two examples use the *same* numbers and the *same* key words. In one problem, you must add to find the answer, while in the other, you must subtract.

EXAMPLE 1 Judy bought 4 cans of pineapple and 16 cans of applesauce. What was the total number of cans that she bought?

STEP 1 *question:* What was the total number of cans?

STEP 2 *necessary information:* 4 cans of pineapple, 16 cans of applesauce.

$$\begin{array}{r} 16 \\ + 4 \\ \hline 20 \end{array}$$

STEP 3 *key words:* and, total
Since you are looking for a total, you should add.

STEP 4 4 cans of pineapple + 16 cans of applesauce = **20 cans of fruit**

EXAMPLE 2 Judy bought a total of 16 cans of fruit. Four were cans of pineapple. The rest were applesauce. How many cans of applesauce did she buy?

STEP 1 *question:* How many cans of applesauce did she buy?

STEP 2 *necessary information:* 4 cans of pineapple, 16 cans of fruit

$$\begin{array}{r} 16 \\ -\ 4 \\ \hline 12 \end{array}$$

STEP 3 *key word:* total
Since you have been given a total and are being asked to find a part of it, you must subtract.

STEP 4 16 cans of fruit − 4 cans of pineapple = **12 cans of applesauce**

In both examples, the word *total* was used. In Example 1, the question asked you to find the total. Therefore, you had to add. But in Example 2, the total (cans of fruit) was part of the information given in the problem. The question asked you to find the number of cans of applesauce, a part of the total. To do this, you had to subtract the number of cans of pineapple from the total number of cans.

These two examples show that key words can be good clues, **but they are only a guide to understanding a word problem.** If you use the key words without understanding what you are reading, you may do the wrong arithmetic.

· ·

This exercise will help you to carefully examine problems containing key words. In each of the following items, the key word has been left out and the solution has been given. Two choices have been given for the missing word; circle the correct one.

1. Last week eggs cost 87 cents a dozen. This week the price _____ 9 cents. How much are eggs this week?

   87 cents + 9 cents = **96 cents**   (fell, rose)

2. Last week the price of eggs _____ to 87 cents a dozen. The price had originally been 78 cents. By how much did the eggs change in price?

   87 cents − 78 cents = **9 cents**   (fell, rose)

3. The 5% sales tax is going to _____ 1%. What will the new sales tax be?

   5% + 1% = **6%**   (increase, decrease)

4. The 5% sales tax is going to _____ 1%. What will the new sales tax be?

   5% − 1% = **4%**   (increase, decrease)

5. Next month, the Jones family is going to receive $14 a week _____ for food stamps. They now receive $87 a week. How much a week will they be receiving?

   $87 − $14 = **$73**   (more, less)

6. The Johnson family's food stamp allotment has been cut. They now receive $14 a week _____, or $87 for stamps. What had been their original allotment for stamps?

   $87 + $14 = **$101**   (more, less)

7. Gloria used to keep her thermostat at 72 degrees. To save energy, she _____ it 6 degrees. What was the new temperature in her apartment?

   72 degrees − 6 degrees = **66 degrees**   (raised, lowered)

8. Jerline's mother came to visit for the weekend. To make sure that her mother was comfortable, she _____ the thermostat to 70 degrees. Usually, the thermostat is set at 66 degrees. By how much has Jerline changed the temperature?

   70 degrees − 66 degrees = **4 degrees**   (lowered, raised)

9. Gail normally ate 2,400 calories a day. While on a special diet, she ate 1,100 calories _____. How many calories a day did she eat on her diet?

   2,400 calories + 1,100 calories = **3,500 calories**   (more, less)

Restating the Problem

Have you ever tried to help someone else work out a word problem? Think about what you do. Often, you read the problem with the person, then discuss it or put it in your own words to help the person see what is happening. You can use this method—**restating the problem**—to help yourself solve a problem.

Restating the problem can be especially helpful when the word problem contains no key words. Look at the following example:

EXAMPLE Susan has already driven her car 2,700 miles since its last oil change. She still plans to drive 600 miles before changing the oil. How many miles does she plan to drive between oil changes?

STEP 1 *question:* How many miles does she plan to drive between oil changes?

STEP 2 *necessary information:* 2,700 miles, 600 miles

STEP 3 Decide what arithmetic operation to use. Restate the problem in your own words: "You are given the number of miles Susan has already driven and the number of miles more that she plans to drive. You need to find the total number of miles between oil changes. You should add."

STEP 4 2,700 miles + 600 miles = **3,300 miles** between oil changes

$$\begin{array}{r} 2{,}700 \\ +600 \\ \hline 3{,}300 \end{array}$$

STEP 5 It makes sense that she will drive 3,300 miles between oil changes since you are looking for a number larger than the 2,700 miles that she has already driven.

Try this method with the next exercise. Read the problem, and then restate it to yourself. In future work with particularly confusing word problems, you should try this method of talking to yourself to understand the problems.

Each word problem is followed by two short explanations. One gives you a reason to add to find the answer. The other gives you a reason to subtract to find the answer. Circle the correct explanation. DO NOT SOLVE!

1. Margi's weekly food budget has increased $12 over last year's to $87 per week. How much had she spent per week for food last year?

   a. The budget has increased since last year. Therefore, you add the two numbers.
   b. Her food budget has increased over last year's. The new, larger budget is given. Therefore, you subtract to find last year's smaller amount.

2. In the runoff election for mayor, Fritz Neptune got 14,662 votes, and Julio Cortez got 17,139 votes. How many votes were cast in the election?

   a. To find the total number of votes cast, you should add the two numbers given.
   b. To find the number of votes cast, you should subtract to find the difference between the two numbers.

3. The difference between first class (the most expensive fare) and the coach air fare is $68. If coach costs $212, how much does first class cost?

   a. To find the cost of first class, you subtract to find the difference between the two fares.
   b. First class costs more than coach. Since you are looking for the larger fare, you add the smaller fare to the difference between the two fares.

4. It costs $66,840 to run and maintain the town's pool. During the year, $59,176 was collected from user fees for the pool, and the town government paid the rest of the cost. How much money did the town government have to pay?

   a. To find the total cost, you add the cost of running and maintaining the pool to the amount collected in user fees.
   b. You are given the total cost of running the pool and the part of the cost covered by user fees. To find the cost to the town government, you subtract.

5. After reading a 320-page novel, Danyel read a 205-page history book. How many pages did Danyel read?

   a. Since you are looking for the total number of pages, you add.
   b. To find the difference between the number of pages in the two books, you subtract.

6. After making 24 bowls, Claire made 16 plates. How many pieces did she make?

   a. Since the number of pieces includes the number of bowls and plates, you add them together.
   b. Since you are looking for a difference, you subtract the number of plates from the number of bowls.

7. A factory has produced 48,624 microwave ovens so far this year. The company expects to produce 37,716 microwave ovens during the rest of the year. What is the projected production of ovens for the year?

   a. To find the projected production for the year, you subtract the number of microwave ovens to be produced from the number of ovens that have been produced so far.
   b. To find the projected production for the entire year, you add the number of ovens already produced to the number of ovens that are expected to be produced.

8. Diane has a 50,000-mile warranty on her car. The car has gone 34,913 miles. As of today, how many miles will the car have left on its warranty?

   a. To find the total number of miles that the car has left on its warranty, you add the number of miles Diane has driven to the number of miles that the warranty covers.
   b. Since Diane has driven on the warranty, you subtract the miles she has already driven from the mileage that the warranty covers.

Using Pictures and Diagrams to Solve Word Problems

Another approach that people use to solve word problems is to form a picture of the problem. While some people can do this in their heads, many people find it very useful to draw a picture or diagram of the problem.

EXAMPLE 1 A recipe for 48 ounces of punch calls for 23 ounces of fruit juice and liquor. The rest is club soda. How much of the recipe is club soda?

STEP 1 *question:* How much of the recipe is club soda?

STEP 2 *necessary information:* 48 ounces of punch, 23 ounces of fruit juice and liquor

STEP 3 Draw a diagram, and decide whether to add or subtract.

The diagram shows that you can find the remaining contents by subtraction.

punch – fruit juice and liquor = club soda

STEP 4 Do the arithmetic.

48 oz – 23 oz = **25 oz**

STEP 5 Make sure that your answer is sensible. It makes sense that the number of ounces of club soda is less than the number of ounces of punch.

EXAMPLE 2 After losing $237 at the blackjack table, Yolanda had $63 left for spending money for the rest of her vacation. How much spending money had she brought with her?

STEP 1 *question:* How much spending money had she brought with her?

STEP 2 *necessary information:* $237 lost, $63 left

STEP 3 Draw a diagram, and decide whether to add or subtract.

The diagram shows that you can find the total spending money by addition.

$ lost + $ left = total $ spending money

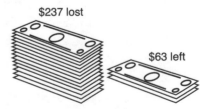

STEP 4 Do the arithmetic.

$237 + $63 = **$300**

STEP 5 Make sure that your answer is sensible. Since she lost money, it makes sense that Yolanda started with more money than she has now.

EXAMPLE 3 The oil tanker *Whyon* was loaded with 150,000 barrels of crude oil when it struck a reef and spilled most of its oil. Within a week, the cleanup crew had pumped all the remaining oil from the tanker. If the cleanup crew pumped 97,416 barrels of oil from the ship, how many barrels of oil were spilled?

STEP 1 *question:* How many barrels of oil were spilled?

STEP 2 *necessary information:* 150,000 barrels of crude oil, 97,416 barrels of crude oil

STEP 3 Draw a diagram, and decide whether to add or subtract.

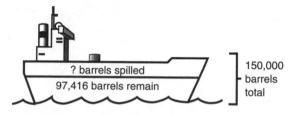

The diagram shows that you can find the amount spilled by subtraction.

total load − amount left = amount spilled

STEP 4 Do the arithmetic.

$$\begin{array}{r} 150{,}000 \text{ barrels} \\ -\ \ 97{,}416 \text{ barrels} \\ \hline \mathbf{52{,}584 \text{ barrels}} \end{array}$$

STEP 5 Make sure that your answer is sensible. It makes sense that the amount spilled is less than the total amount of oil that was originally on the tanker.

⋯⋯⋯⋯⋯⋯⋯⋯⋯⋯⋯⋯⋯⋯⋯⋯⋯⋯⋯⋯⋯⋯⋯⋯

For each problem, make a drawing or a diagram and decide whether to add or subtract. Then solve the problem.

1. If three more students are added to this class, we will have 31 students. How many students do we have now?

2. Rafael Hernandez paid $39 less in taxes in 1998 than in 1999. He paid $483 in 1998. How much did he pay in 1999?

3. Every hour 12,000 gallons of water flow through the dam spillway. The 41-year-old dam operator plans to decrease the flow by 3,500 gallons an hour. What will be the new rate of water flow?

4. A $120 ink-jet printer costs $359 less than a laser printer. How much does the laser printer cost?

5. An ink-jet printer costs $120 less than a $359 laser printer. How much does the ink-jet printer cost?

6. Between 6 P.M. and 11 P.M., the temperature decreased by 13 degrees to 61 degrees. What had the temperature been at 6 P.M.?

7. At 6 P.M. the temperature was 61 degrees. Between 6 P.M. and 11 P.M., it decreased 13 degrees. What was the temperature at 11 P.M.?

8. In 1938 1,412 people graduated from Lincoln High School. Today, 457 of these graduates are still living. How many of the graduates have died?

9. Marion took out a loan for $6,000. She has paid back $3,800. How much does she still owe?

10. A 2,600-pound truck can carry a 1,000-pound load. How much does the fully loaded truck weigh?

11. Gamma Airlines allows each passenger to check a maximum of two bags that together must weigh less than 70 pounds. At check-in, Khanh's bags weighed 24 pounds and 42 pounds. Did the total weight of her bags exceed the weight limit?

12. Macrohard Corporation sold 790,000 copies of their game software Fog. They estimate that there are 600,000 illegal copies of Fog. If their estimates are correct, how many copies of Fog exist?

Using Number Sentences to Solve Word Problems

Addition and subtraction word problems can be solved by writing number sentences. A **number sentence** restates a word problem first in words and then in numbers.

EXAMPLE 1 Lori went to school for 5 years in Levittown before moving to Plainview. She then went to school for 7 years in Plainview. For how many years did she go to school?

To write a number sentence, first write the information in the problem in words.

Levittown plus Plainview equals total years

Then substitute numbers and mathematical symbols for the words.

5 years + 7 years = total years

Solve.

12 years = total years

$$\begin{array}{r} 5 \\ + 7 \\ \hline 12 \end{array}$$

EXAMPLE 2 A play ran for two nights at a theater seating 270 people. The first night 235 people saw the play, and 261 people saw the play the second night. How many people saw the play during its two-night run?

STEP 1 *question:* How many people saw the play during its two-night run?

STEP 2 *necessary information:* 235 people, 261 people

STEP 3 *number sentence:*

first night + second night = total people
235 people + 261 people = total people

$$\begin{array}{r} 235 \\ + 261 \\ \hline 496 \end{array}$$

STEP 4 **496 people = total people**

EXAMPLE 3 Gloria bought a $57 dress on sale for $19. How much did she save?

STEP 1 *question:* How much did she save?

STEP 2 *necessary information:* $57, $19

STEP 3 *number sentence:*

original price − sale price = savings
$57 − $19 = savings

$$\begin{array}{r} 57 \\ - 19 \\ \hline 38 \end{array}$$

STEP 4 **$38 = savings**

Underline the necessary information. Write a word sentence and a number sentence. Then solve the problem.

1. Ross needed a 20-cent stamp. If he paid for the stamp with a quarter, how much change did he get?

2. Bruce drives 32 miles to work each day. When he arrived at work on Monday, he found that he had driven 51 miles that day. How many additional miles over his regular commuting distance had Bruce driven on Monday?

3. The theater company needs to sell 172 Saturday tickets to break even. How many more Saturday tickets must they sell in order to break even according to the chart at the right?

Ticket Sales	
Thursday	120
Friday	145
Saturday	134

4. Wendy decided to buy a $3,300 used car. She had saved $1,460. She got a loan for the rest. What was the amount of the loan?

5. Becci Bachman needs 150 names on her nominating petition to run for office. She collected 119 names on her first day of campaigning. How many more names does she have to collect?

6. After losing 47 pounds, Ann weighed 119. What was her original weight?

7. Lucy's monthly food stamp allotment was reduced by $13 to $168. How much was she getting in food stamps before the reduction?

8. The refrigerator shown at the right was marked down to $379. How much did Kathy save by buying the refrigerator on sale?

$465

9. John had $213 withheld for federal income tax. In fact, he only owed $185. How much of a refund will he receive?

10. A car factory cut production by 3,500 cars to 8,200 cars a month. What had the monthly production been before the cutback?

11. Maria earned $28,682 last year. She spent $27,991. How much did she save?

12. Mr. Crockett's cow Bertha produced 1,423 gallons of milk last year. His other cow, Calico, produced 1,289 gallons. How much milk did his cows produce last year?

13. Memorial Stadium has 72,070 seats. At the football game, 58,682 people had seats. How many seats were empty?

14. In one garden bed, a gardener grew spinach. When the spinach was harvested, he grew green beans. The spinach was harvested after 49 days. The green beans were harvested after 56 days. For how many days were vegetables growing in the garden bed?

ADDITION AND SUBTRACTION WORD PROBLEMS: DECIMALS AND FRACTIONS

Using the Substitution Method

So far, you have solved addition and subtraction word problems using whole numbers. However, many students worry when they see word problems using large whole numbers, fractions, or decimals.

Read the following examples and think about their differences and similarities.

EXAMPLE 1 A cardboard manufacturer makes cardboard 4 millimeters thick. To save money, he plans to make cardboard 3 millimeters thick instead. How much thinner is the new cardboard?

STEP 1 *question:* How much thinner is the new cardboard?

STEP 2 *necessary information:* 4 mm, 3 mm

STEP 3 Decide what arithmetic operation to use. You are given the thickness of each piece of cardboard. Since you must find the difference between the two pieces, you should subtract.

$$\begin{array}{r} 4 \\ -\ 3 \\ \hline 1 \end{array}$$

STEP 4 4 mm – 3 mm = **1 mm**

> **Note:** *mm* stands for *millimeter.* You should be able to do this type of problem even if you aren't familiar with the units of measurement.

EXAMPLE 2 A cardboard manufacturer makes cardboard 6.45 millimeters thick. To save money, he plans to make cardboard 5.5 millimeters thick instead. How much thinner is the new cardboard?

STEP 1 *question:* How much thinner is the new cardboard?

STEP 2 *necessary information:* 6.45 mm, 5.5 mm

STEP 3 Decide what arithmetic operation to use. You are given the thickness of each piece of cardboard. Since you must find the difference between the two pieces, you should subtract. Be sure to put the decimal points one under the other.

$$\begin{array}{r} 6.45 \\ -\ 5.50 \\ \hline 0.95 \end{array}$$

STEP 4 6.45 mm – 5.50 mm = **0.95 mm**

EXAMPLE 3 A cardboard manufacturer makes cardboard $\frac{3}{8}$ inch thick. To save money, he plans to make cardboard $\frac{1}{3}$ inch thick instead. How much thinner is the new cardboard?

STEP 1 *question:* How much thinner is the new cardboard?

STEP 2 *necessary information:* $\frac{3}{8}$ inch, $\frac{1}{3}$ inch

STEP 3 Decide what arithmetic operation to use. You are given the thickness of each piece of cardboard. Since you must find the difference between the two pieces, you should subtract.

$$\begin{array}{r} \frac{3}{8} = \frac{9}{24} \\ -\frac{1}{3} = \frac{8}{24} \\ \hline \frac{1}{24} \end{array}$$

STEP 4 $\frac{3}{8}$ in. $- \frac{1}{3}$ in. $= \frac{1}{24}$ **in.**

> **Note:** Remember, to find the solution, you must change unlike fractions to fractions with a common denominator.

What did you notice about the three example problems?

The wording of all three is exactly the same. Only the numbers and labels have been changed. All three problems are solved the same way, by subtracting.

Then why do Examples 2 and 3 seem harder than the first example?

The difficulty has to do with **math intuition,** or the feel that a person has for numbers. You have a very clear idea of the correct answer to $4 - 3$. It is more difficult to picture $7,483,251 + 29,983$ or $6.45 - 5.5$. And for most of us, our intuition totally breaks down for $\frac{3}{8} - \frac{1}{3}$.

Changing only the numbers in a word problem does not change what must be done to solve the problem. By substituting small whole numbers in a problem, you can understand the problem and how to solve it.

EXAMPLE 4 A floor is to be covered with a layer of $\frac{3}{4}$-inch fiberboard and $\frac{7}{16}$-inch plywood. By how much will the floor level be raised?

Fractions, especially those with different denominators, are especially hard to picture. You can make the problem easier to understand by substituting small whole numbers for the fractions. You can substitute any numbers, but try to use numbers under 10. These numbers do not have to look like the numbers they are replacing.

In Example 4, try substituting 3 for $\frac{3}{4}$ and 2 for $\frac{7}{16}$. The problem now looks like this:

> A floor is to be covered by a layer of 3-inch fiberboard and 2-inch plywood. By how much will the floor level be raised?

You can now read this problem and know that you must add.

Once you make your decision about *how* to solve the problem, you can return the original numbers to the word problem and work out the solution. With the substituted numbers, you decided to *add* 3 and 2. Therefore, in the original, you must *add* $\frac{3}{4}$ and $\frac{7}{16}$.

$$\frac{3}{4} = \frac{12}{16}$$
$$+\frac{7}{16} = \frac{7}{16}$$
$$\overline{\quad\quad}$$
$$\frac{19}{16} = 1\frac{3}{16} \text{ inches}$$

> **Remember:** Choosing the whole numbers 3 and 2 was completely arbitrary. You could have used any small whole numbers.

Below is a set of six substitutions. Each of the substitutions will fit only one of the following six word problems. Match the letter of the correct substitution to each problem. (After each problem, you are told which numbers to substitute for that problem. To keep it simple, we are only using the numbers 4, 3, and 1 in the substitutions.)

> **a.** 3 pounds + 4 pounds = 7 pounds
> **b.** 4 pounds − 3 pounds = 1 pound
> **c.** $3 − $1 = $2
> **d.** $3 + $1 = $4
> **e.** 3 inches + 1 inch = 4 inches
> **f.** 3 inches − 1 inch = 2 inches

1. A sweater that normally sells for $35.99 has been marked down by $10.99. What is the sale price of the sweater?

   Substitute $3 for $35.99 and $1 for $10.99.

   _____

2. How much heavier is the rump roast than the round roast shown at the right?

   Substitute 3 for 3.46 and 4 for 4.17.

   _____

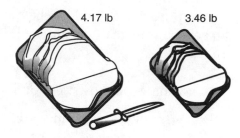

4.17 lb   3.46 lb

3. Robin bought a 3.28-pound steak and a 4.84-pound chicken. What was the weight of the meat she bought?

   Substitute 3 for 3.28 and 4 for 4.84.

   _____

4. Janice bought a skirt for $31.99 and a slip for $11.59. How much did she spend in all?

   Substitute $3 for $31.99 and $1 for $11.59.

   _____

5. Michael caught a $21\frac{1}{4}$-inch fish. His friend Paul caught a $23\frac{1}{16}$-inch fish. How much longer was Paul's fish?

   Substitute 1 for $21\frac{1}{4}$ and 3 for $23\frac{1}{16}$.

   _____

6. Two boards were placed end-to-end. The first board was $40\frac{7}{8}$ inches long. The second board was $32\frac{3}{4}$ inches long. What was the combined length of the 2 boards?

   Substitute 3 for $40\frac{7}{8}$ and 1 for $32\frac{3}{4}$.

   _____

Using Estimation

When your car is in an accident and you take it to an auto body shop for repairs, you first receive an **estimate** for the cost of the repairs. This might not be the exact or final price, but it should be close.

When solving word problems, it is also important to have some idea of what the answer should be before you start doing the arithmetic. You can get an estimate of the answer by approximating the numbers in the problem.

An **estimate** is almost, but not quite, the exact number. For instance,

> In the last election, the newspaper reported that Alderman Jones received 52% of the vote and his opponent received 48%. Actually, the alderman received 52.1645% of the vote and his opponent received 47.8355%.

The newspaper did not report the exact percent of the vote; it **rounded** the numbers to the nearest whole percent. Rounded numbers are one type of estimate.

Estimating the numbers and doing quick arithmetic in your head is a good way to check your work. Throughout the rest of this workbook, you should first estimate a solution before doing the calculation. The estimation icon will indicate that the problem is best solved by rounding and estimating.

 Below is a set of six estimated solutions. The numbers in the solutions have been rounded. Match each word problem with one of the solutions.

a. 9 miles − 7 miles = 2 miles
b. 9 miles + 7 miles = 16 miles
c. 4 billion dollars − 1 billion dollars = 3 billion dollars
d. 4 billion dollars + 1 billion dollars = 5 billion dollars
e. 37,000 fans − 35,000 fans = 2,000 fans
f. 37,000 fans + 35,000 fans = 72,000 fans

1. During the last weekend in July, 35,142 fans saw a baseball game on Saturday. On Sunday 36,994 fans saw a game. What was the total attendance for the weekend?

   _____

2. Nationwide, Grand Discount stores sold 37,238 window fans in April and 34,982 fans in May. How many more fans were sold in April?

   _____

3. The original estimate for the cost of a nuclear power plant was .984 billion dollars. The final cost was 4.16 billion dollars. How much did the price increase from the original estimate?

   _____

4. The state budget is 3.92 billion dollars. It is expected to increase 1.2 billion dollars over the next 5 years. How much is the budget expected to be 5 years from now?

   _____

5. By expressway, it is $7\frac{1}{4}$ miles to the beach. By back roads, it is $8\frac{9}{10}$ miles. How much shorter is the trip when driving by expressway?

   _____

6. Pat is a long-distance runner. He ran $6\frac{9}{10}$ miles on Saturday and $9\frac{1}{8}$ miles on Sunday. How many miles in all did he run during the weekend?

   _____

Decimals: Restating the Problem

Restating the problem is one method that will work as well with solving decimal problems as with whole-number problems. Don't worry about the decimal points until after you have decided to add or subtract. Then, remember to line up the decimal points before doing the arithmetic.

EXAMPLE   A pair of pants was on sale for $8.99. A shirt was on sale for $6.49. Alan decided to buy both. How much did he spend?

STEP 1   *question:* How much did he spend?

STEP 2   *necessary information:* $8.99, $6.49

STEP 3   *restatement:* Since Alan is buying both items, you add to find the total amount he spent.

$$\begin{array}{r} 8.99 \\ + \ 6.49 \\ \hline 15.48 \end{array}$$

STEP 4 $8.99 + $6.49 = **$15.48**

STEP 5   Round $8.99 to $9 and $6.49 to $6.

$9 + $6 = $15

Therefore, your answer should be close to $15. Making an estimate is a good method of checking your answer and making sure it is sensible.

Circle the letter of the correct restatement and solve the problem. Use estimation to make sure your answer is sensible.

1. Using his odometer, George discovered that one route to work was 6.3 miles long and the other was 7.1 miles. How much shorter was the first way?

    a. Since you are given the two distances to work, add to find out how much shorter the first way was.
    b. To find how much shorter the first way was, subtract to find the difference.

2. Max had to put gasoline in his 8-year-old car twice last week. The first time, he put in 9.4 gallons. The second time, he put in 14.7 gallons. How much gasoline did he put in his car last week?

   a. To find the total amount of gasoline he put in his car, you add.
   b. Since you are given the two amounts of gasoline, you subtract to find the difference.

3. The first fish fillet weighed 1.42 pounds. The second fillet weighed 0.98 pound. Alice decided to buy both fillets. What was the weight of the fish she bought?

   a. To find the total weight of the two fish fillets, you add.
   b. Since you are given the weight of the two fish fillets, you subtract to find the difference between their weights.

4. At the Reckless Speedway, Bobby was clocked at 198.7 mph, while Mario was clocked at 200.15 mph. How much faster did Mario drive than Bobby?

   a. Add the two speeds to find how much faster Mario drove.
   b. Since Mario drove faster, subtract Bobby's speed from his to find out the difference between the speeds.

5. Last year the unemployment rate was 7.9%. This year it has increased to 9.1%. By how much did unemployment rise?

   a. To find how much unemployment rose, add the two unemployment rates.
   b. To find the rise in unemployment, subtract last year's rate from this year's rate.

Decimals: Drawings and Diagrams

Diagrams and drawings can help you solve decimal addition or subtraction word problems.

EXAMPLE   A metal bearing was 0.24 centimeters thick. The machinist ground it down until it was 0.065 centimeters thinner. How thick was the metal bearing after it had been ground down?

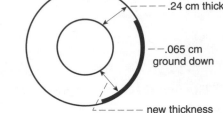

- .24 cm thick
- .065 cm ground down
- new thickness

STEP 1   *question:* How thick was the metal bearing after it had been ground down?

STEP 2   *necessary information:* 0.24 cm, 0.065 cm

STEP 3   *make a drawing:* To find the size of the bearing after it was ground down, you subtract.

STEP 4   Do the arithmetic. Be sure to line up the decimal points. If you add a zero, you can see that 0.24 (0.240) is larger than 0.065.

$$\begin{array}{r} 0.240 \\ -\ 0.065 \\ \hline 0.175 \end{array}$$

0.240 cm – 0.065 cm = **0.175 cm**

> **Remember:** When subtracting decimals, first line up the decimal points. Then fill any blank spaces to the right of the decimal point with zeros. This should help you borrow correctly.

Make a drawing or a diagram, and solve the problem. (Each person's drawing may be different. What is important is that the diagram makes sense to you.)

1. Meatball subs used to cost $2.60 at Mike's, but he just raised the price $0.25. How much do meatball subs cost now?

2. Tara's prescription for 0.55 gram of antibiotic was not strong enough. Her doctor gave her a new prescription for 0.7 gram of antibiotic. How much stronger was the new prescription?

3. Mike Johnson was hitting .342 before he went into a batting slump. By the end of his slump, his average had dropped .083. What was his batting average at the end of his slump?

4. Joyce earned $313.50 and had $126.13 taken out for deductions. How much was her take-home pay?

5. A wooden peg is 1.6 inches wide and 3.2 inches long. It can be squeezed into an opening 0.05 inch smaller than the width of the peg. What is the width of the opening?

6. By midweek Wendy had spent $46.65. At the end of the week, she had spent $23.35 more. How much did Wendy spend that week?

7. The gap of a spark plug should be 0.08 inch. The plug would still work if the gap were off by as much as 0.015 inch. What is the largest gap that would still work?

8. The King Coal Company mined 126.4 tons of coal. Because of high sulfur content, 18.64 tons of coal were unusable. How many tons of coal were usable?

9. Bonnie was mixing chemicals in a lab. The formula called for 1.45 milliliters of sulfuric acid, but she had 1.8 milliliters of sulfuric acid in her pipette. How much extra sulfuric acid does she have in the pipette? (A pipette is a glass tube used for measuring chemicals.)

10. Barbara complained that the 2.64-pound steak had too much excess fat. The butcher trimmed the steak and reweighed it. It now weighed 2.1 pounds. How much fat did the butcher cut off the steak?

Decimals: Writing Number Sentences

Number sentences can help you solve decimal addition and subtraction word problems. Look at the following examples to see how number sentences are used.

EXAMPLE 1 Meryl bought $16.27 worth of groceries and paid with a $20 bill. How much change did she receive?

STEP 1 *question:* How much change did she receive?

STEP 2 *necessary information:* $16.27, $20

STEP 3 *number sentence:*

amount paid – price of groceries = change
$20.00 – $16.27 = change

$$\begin{array}{r} \$20.00 \\ -\$16.27 \\ \hline \$\ 3.73 \end{array}$$

STEP 4 **$3.73 = change**

EXAMPLE 2 On sale a pair of pants costs $22.49. They had been discounted $4.49 from the original price. What was the original price?

STEP 1 *question:* What was the original price?

STEP 2 *necessary information:* $4.49, $22.49

STEP 3 *number sentence:*

sale price + discount = original price
$22.49 + $4.49 = original price

$$\begin{array}{r} \$22.49 \\ +\$\ 4.49 \\ \hline \$26.98 \end{array}$$

STEP 4 **$26.98 = original price**

Underline the necessary information. Write a word sentence and a number sentence. Then solve the problem.

1. The Sticky Candy Company decided to reduce the size of their chocolate candy bar by 0.6 ounce to 2.4 ounces. How much did their chocolate bar weigh before the change?

2. Julie's lunch cost $3.38. If she paid with a $10 bill, how much change did she get?

3. Bernice bought one chicken that weighed 3.94 pounds and one that weighed 4.68 pounds. She also bought a 1.32-pound steak. How much chicken did she buy?

4. The odometer at the right shows Mark's car mileage when he left Boston. When he arrived in New York, the odometer read 23,391.4 miles. How long was the trip?

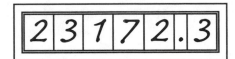

5. The chart at the right shows the costs of subway and bus rides in Connie's city. If Connie needs to take one bus and one subway to her mother's house, how much will it cost her for a one-way trip?

Fares	
Bus	$.40
Subway	$.65

6. Judy spent $341.98 for a new washing machine in Massachusetts. If she had bought the same machine in New Hampshire, she would have paid $335.26, since that state does not have a sales tax. How much less would she have paid in New Hampshire?

7. When he ran the 200-meter race, Marcus ran the first 100 meters in 14.36 seconds and the second 100 meters in 13.9 seconds. What was his total time for the race?

8. One assembly line at the plant produced 966 soda bottles in one hour. Another line produced 50 fewer bottles in the same amount of time. How many bottles did the second line produce in an hour?

Fractions: Restating the Problem

In this section, you will restate the problem in order to decide whether to add or subtract. Before solving these fraction problems, you might want to use estimation to help you decide what arithmetic operation to use.

EXAMPLE   Tanya grew $2\frac{3}{4}$ inches last year. If she was $42\frac{1}{2}$ inches tall a year ago, how tall is she now?

    STEP 1   *question:* How tall is she now?

    STEP 2   *necessary information:* $2\frac{3}{4}$ inches, $42\frac{1}{2}$ inches

    STEP 3   *restatement:* Since you know Tanya's old height, and you know that she grew, you must add to find her new height.

    STEP 4   *estimation:* 3 inches + 43 inches = 46 inches

    STEP 5 $2\frac{3}{4}$ inches + $42\frac{1}{2}$ inches = height now

               $2\frac{3}{4} + 42\frac{2}{4} = 44\frac{5}{4}$ inches = **$45\frac{1}{4}$ inches now**

$$2\frac{3}{4} = 2\frac{3}{4}$$
$$+ 42\frac{1}{2} = 42\frac{2}{4}$$
$$\rule{3cm}{0.4pt}$$
$$44\frac{5}{4} = 45\frac{1}{4}$$

> **Remember:** Whenever you add or subtract fractions, find a common denominator.

Each problem is followed by two restatements and estimations. Circle the correct restatement. Then solve the problem.

1. A carpenter needed one piece of molding $28\frac{1}{2}$ inches long and a second piece $31\frac{1}{4}$ inches long. How much molding did he need?

    a. To find out how much molding is needed, you should subtract.

        31 inches – 29 inches = 2 inches

    b. To find the total amount of molding needed, you should add.

        29 inches + 31 inches = 60 inches

2. Vera combined $1\frac{2}{3}$ cups of flour and $1\frac{1}{3}$ cups of butter in a 2-quart mixing bowl. How many cups of the mixture did Vera have?

   a. Since Vera is combining the flour and butter, the amount of the mixture can be found by adding.

      2 cups + 1 cup = 3 cups

   b. Since you are given the amount of flour and the amount of butter, you subtract to find the amount of the mixture.

      2 cups − 1 cup = 1 cup

3. According to the scales at the right, how much heavier is Tara than Erin?

   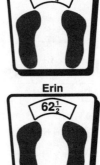

   a. Since you are comparing two weights, subtract to find the difference.

      71 pounds − 63 pounds = 8 pounds

   b. Since you are finding Tara's total weight, add the given weights.

      71 pounds + 63 pounds = 134 pounds

4. Mira was $18\frac{3}{4}$ inches tall at birth. Six months later, she was $23\frac{1}{4}$ inches tall. How much taller was Mira after 6 months than at birth?

   a. To find how much taller Mira is, add the given heights.

      19 inches + 23 inches = 42 inches

   b. To find how much taller Mira is, subtract her birth height from her height at 6 months.

      23 inches − 19 inches = 4 inches

Fractions: Diagrams and Pictures

Making diagrams and pictures can also help you solve fraction addition or subtraction word problems.

EXAMPLE Using a $\frac{3}{8}$-inch drill bit, Judy drilled a hole that was slightly too small. She used the next size drill bit, one that was $\frac{1}{32}$ inch larger, to enlarge the hole. What was the size of the new drill bit?

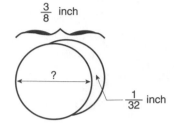

$\frac{3}{8}$ inch

$\frac{1}{32}$ inch

STEP 1 *question:* What was the size of the new drill bit?

STEP 2 *necessary information:* $\frac{3}{8}$ inch, $\frac{1}{32}$ inch

STEP 3 Decide what arithmetic operation to use. Draw a picture. Since you are looking for the next larger size, you should add.

STEP 4 $\frac{3}{8}$ inch $+ \frac{1}{32}$ inch $= \frac{12}{32}$ inch $+ \frac{1}{32}$ inch $= \frac{13}{32}$ **inch**

$$\frac{3}{8} = \frac{12}{32}$$
$$+\frac{1}{32} = \frac{1}{32}$$
$$\frac{13}{32}$$

Make a drawing or diagram and solve each of the problems below.

1. A piece of wood called a 2-by-4 (a 2-inch by 4-inch board) is really not 4 inches wide. It is actually $\frac{5}{8}$ inch narrower. What is the real width of the board?

2. Pat is at the hospital for a total of $8\frac{1}{2}$ hours a day. If during each day he has $1\frac{3}{4}$ hours for breaks, how long does he work each day?

3. Hope worked $6\frac{1}{2}$ hours and took an additional $\frac{3}{4}$ hour for lunch. What was the total amount of time that Hope spent at work and lunch?

4. A 2-by-4 is not really 2 inches thick. It is $\frac{1}{2}$ inch thinner. What is the real thickness of the board?

5. A recipe called for $2\frac{1}{2}$ cups of flour. George only had $1\frac{2}{3}$ cups. How much flour did he borrow from his neighbor?

6. Felix tried to loosen a bolt with a $\frac{3}{4}$-inch wrench, but it was slightly too large. He decided to try the next smaller size, which was $\frac{1}{16}$ inch smaller. What was the size of the next smaller size wrench?

Fractions: Using Number Sentences

Number sentences can also help you solve fraction addition or subtraction word problems.

EXAMPLE   David planned to make a 3-inch-thick insulated roof. The roof will be made with a layer of thermal board on top of $\frac{5}{8}$-inch plywood. How thick can the thermal board be?

STEP 1   *question:* How thick can the thermal board be?

STEP 2   *necessary information:* 3 inches, $\frac{5}{8}$ inch

STEP 3   *number sentence:*

thickness of roof – plywood = thermal board

3 inches – $\frac{5}{8}$ inch = thermal board

STEP 4 $2\frac{3}{8}$ **inches = thermal board**

$$3 = 2\frac{8}{8}$$
$$-\frac{5}{8} = \frac{5}{8}$$
$$\overline{2\frac{3}{8}}$$

Write a word sentence and a number sentence for each problem. Then solve the word problems.

1. Beverly filled a 3-quart punch bowl with punch. If she used $1\frac{1}{4}$ quarts of rum, how many quarts of other ingredients did she use?

2. Amy bought a skirt that was the length shown below. If she shortened it to $32\frac{3}{4}$ inches, how much did she take off?

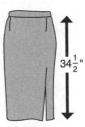

$34\frac{1}{2}$"

3. After 3 weeks in the store, a bolt of cloth that had originally been 20 yards long was $6\frac{1}{2}$ yards long. Then $3\frac{2}{3}$ more yards of the cloth were sold. How much cloth was left?

4. Last winter Fred used $\frac{1}{8}$ cord of wood one week and $\frac{1}{12}$ cord of wood the next week to heat his house. How much wood did he use during the 2 weeks?

5. Linda bought $62\frac{1}{2}$ inches of cloth to make drapes. She used $\frac{3}{4}$ inch for the hem. How long were the drapes?

Using Algebra to Solve Word Problems

You have already seen how to use number sentences to solve addition and subtraction word problems. In algebra, instead of using words, you use a letter of the alphabet to stand for the number you are looking for.

A **number sentence** in which one amount is equal to another amount is called an **equation.** To find what number will make an equation true, you must solve the equation. One way to solve an equation is shown in the following example.

EXAMPLE   A cup of 2% milk contains 130 calories, of which 45 calories are from fat. How many calories in the cup of 2% milk are not from fat?

STEP 1   *question:* How many calories in the cup of 2% milk are not from fat?

STEP 2   *necessary information:* 130 calories, 45 calories

STEP 3   *number sentence or equation:*

calories from fat + calories not from fat = total calories
45 calories + calories not from fat = 130 calories

You can rewrite the number sentence using the letter *c* to stand for *calories not from fat.* (You can select any letter, but *c* is a good choice since it is the first letter of *calories.*) 45 calories + *c* = 130 calories

STEP 4   *solve the equation:*

One way to solve the equation is to subtract 45 calories from each side so that *c* stands alone on the left side of the equation. You can do this because if you subtract the same amount from both sides of an equation, the results are still equal.

$$
\begin{array}{rcr}
45\ \text{calories} + c & = & 130\ \text{calories} \\
-\ 45\ \text{calories} & = & -\ 45\ \text{calories} \\
\hline
c & = & 85\ \text{calories}
\end{array}
$$

STEP 5   *Does the answer make sense?*

45 calories + 85 calories = 130 calories. The answer makes sense.

Estimation: The letter is a placeholder for a number. You need to find the number that will make the equation true. To build your math intuition, try to substitute a number for the letter before you start the formal solution. Try to get a good estimate of the answer before doing the detailed calculations.

**Write a word sentence and an equation for each problem. Then solve
the word problem.**

1. In January the adult education student Website on the Internet had
   2,917 hits or visits. On February the site had 4,348 hits. How many
   more hits did the site have in February?

2. Belquis has a new home computer and went shopping for
   educational software for her son, Jonathan. She bought Math
   Mania for $9.98 and paid for it with a $20 bill. What was her
   change?

3. A slice of cheese pizza has 134 calories. A slice of pepperoni pizza
   has 149 calories. How many calories are in the pepperoni on a
   pepperoni pizza?

4. In order to patch a water-damaged post, Raymonde needed a
   $1\frac{3}{4}$ foot long 1″ × 6″ board. In her basement, she found a $3\frac{1}{2}$ foot
   long 1″ × 6″ board. After she cut the board to make the repair, how
   much was left over?

5. A tablet of Excedrin contains 250 milligrams of acetaminophen,
   250 milligrams of aspirin, and 65 milligrams of caffeine as active
   ingredients. How many milligrams of active ingredients are in a
   tablet of Excedrin?

6. In the fleece mill, inspector Diogenes found that the pile height of a
   batch of fabric was $\frac{5}{8}$ inch. The standard pile height for the fabric
   was $\frac{7}{16}$ inch. How much material needs to be sheared off for the
   fabric to meet the standard?

Solving Addition and Subtraction Word Problems

Solve each problem and circle the letter of the correct answer. If the correct answer is not one of the choices, circle e. none of the above.

1. The Hammerhead Nail Company produces 55,572 nails and 4,186 screws a day. On Monday 1,263 nails were no good. How many good nails were made on Monday?

   a. 56,835 nails
   b. 54,309 nails
   c. 59,758 nails
   d. 51,386 nails
   e. 58,495 nails

2. Ana Marie had $75.62 in her wallet. How much money did she have after spending $38.56?

   a. $114.18
   b. $37.06
   c. $1.98
   d. $71.76
   e. none of the above

3. The gauges at the right show Ron's mileage using different types of gas. How much better was his mileage when he used gasohol?

   a. 4.13 miles
   b. 29 miles
   c. 41.3 miles per gallon
   d. 2.9 miles per gallon
   e. 1.15 miles per gallon

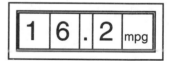

Gasoline

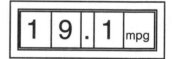

Gasohol

4. When Kathy bought her car, she paid $1,800 down and had $640 left in her savings account. She then paid $6,400 over the next 2 years to finish paying for the car. How much did the car cost her?

   a. $8,840
   b. $4,600
   c. $4,100
   d. $8,200
   e. $7,560

5. Before cooking, a hamburger weighed $\frac{1}{4}$ pound. After cooking, it weighed $\frac{3}{16}$ pound. The rest of the hamburger was fat that burned off during cooking. How much fat burned off during cooking?

    a. $\frac{4}{20}$ pound

    b. $\frac{1}{6}$ pound

    c. $\frac{1}{16}$ pound

    d. $\frac{7}{16}$ pound

    e. none of the above

6. Brand X contains 0.47 gram of pain reliever per 1.5-gram tablet. Brand Y contains 0.6 gram of pain reliever. How much more pain reliever does Brand Y have than Brand X?

    a. 0.53 gram

    b. 0.41 gram

    c. 0.13 gram

    d. 0.27 gram

    e. 2.57 grams

7. A public television station has already raised $391,445 and must raise $528,555 more to stay in business. What was the target amount for the station's fund-raising drive?

    a. $920,000

    b. $137,110

    c. $127,110

    d. $237,110

    e. none of the above

8. Joe bought the window shade at the right. When he got home, he found out that it was $2\frac{3}{8}$ inches too narrow. What size window shade does he need?

    a. $24\frac{3}{8}$ inches

    b. $29\frac{1}{8}$ inches

    c. $28\frac{1}{2}$ inches

    d. 24 inches

    e. none of the above

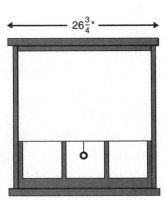

9. To heat her house last winter, Mrs. George used $\frac{5}{8}$ cord of wood in February and $\frac{1}{12}$ cord of wood in March. How much wood did she use?

   a. $\frac{3}{10}$ cord

   b. $\frac{13}{24}$ cord

   c. $\frac{1}{2}$ cord

   d. $\frac{3}{4}$ cord

   e. none of the above

10. A gallon of economy paint contained 3.4 tubes of pigment per gallon. The high-quality paint contained 5.15 tubes of pigment per gallon. What was the difference between the amount of pigment used for each paint?

   a. 4.81 tubes

   b. 5.49 tubes

   c. 8.55 tubes

   d. 2.35 tubes

   e. none of the above

11. A radioactive tracer lost $\frac{1}{2}$ of its radioactivity in an hour. Three hours later it had lost another $\frac{7}{16}$ of its radioactivity. What was the total loss in radioactivity for the entire time?

   a. $\frac{7}{32}$ of its radioactivity

   b. $\frac{15}{16}$ of its radioactivity

   c. $\frac{4}{9}$ of its radioactivity

   d. $\frac{1}{16}$ of its radioactivity

   e. none of the above

12. A gypsy moth grew 0.03 gram from the size shown at the right. How much did the gypsy moth weigh after growing?

   a. 2.08 grams

   b. 2.74 grams

   c. 3.10 grams

   d. 2.80 grams

   e. none of the above

Gypsy Moth

2.77 grams

MULTIPLICATION AND DIVISION WORD PROBLEMS: WHOLE NUMBERS

Identifying Multiplication Key Words

In arithmetic, there are four basic operations: addition, subtraction, multiplication, and division. As you have seen, subtraction can be thought of as the opposite of addition. In the same way, division can be thought of as the opposite of multiplication. This concept is useful in deciding whether a problem is a multiplication or a division problem.

In previous chapters, you looked at addition and subtraction key words. There are also multiplication key words.

EXAMPLE 1   Diane always bets $2 on a race. Last night she bet eight times. How much money did she bet?

multiplication key word: times

EXAMPLE 2   It cost Fernando $39 per day to rent a car. He rented a car for 4 days. How much did he pay to rent the car?

multiplication key word: per

> **Remember:** *Per* means "for each."

Multiplication can also be considered repeated addition. Therefore, it is possible for an addition key word to also be a multiplication key word. *Total* is a word that can indicate either addition or multiplication.

In the following problems, circle the multiplication key words. DO NOT SOLVE!

1. Miguel pays his landlord $670 for rent 12 times a year. How much rent does he pay in a year?

2. During the 9 months that she stayed in her apartment, Isabelle paid $43 per month for electricity. How much did she pay for electricity during the time she stayed in her apartment?

3. At the stable, one horse eats 3 pounds of hay a day. What is the total amount of hay needed to feed 26 horses?

4. When her children were young, Alzette had a part-time job for 18 hours a week. She now works twice as many hours as she did then. How many hours a week does she work now?

5. Sam and Marion bought a new home on an 80-by-90 foot lot. How large, in square feet, was the lot?

6. Amy's living room is 24 feet long and 16 feet wide. What is the area of her living room?

7. Sally's research found that every dollar invested in campaign fund-raising was multiplied eight times by new contributions. If her research is correct, how much in new contributions can she expect if she invests $3,600 in fundraising?

8. Fresh orange juice has 14 calories per ounce. How many calories are in an 8-ounce serving of orange juice?

9. Margo's family drinks 3 gallons of milk per week. At $1.75 per gallon, how much does Margo spend each week for milk?

Solving Multiplication Word Problems with Key Words

Look at the following examples of multiplication key words.

EXAMPLE 1 Shirley cleans the kitchen sink three times a week. How many times does she clean the sink in 4 weeks?

>   **STEP 1** *question:* How many times does she clean the sink?
>
>   **STEP 2** *necessary information:* 3 times a week, 4 weeks
>
>   **STEP 3** Decide what arithmetic operation to use.
>
>   *multiplication key word:* times
>
>   **STEP 4** 3 times a week × 4 weeks = **12 times**

$$\begin{array}{r} 3 \\ \times\ 4 \\ \hline 12 \end{array}$$

EXAMPLE 2 During the Depression, eggs cost 14 cents per dozen. How much did 5 dozen eggs cost?

>   **STEP 1** *question:* How much did 5 dozen eggs cost?
>
>   **STEP 2** *necessary information:* 14 cents per dozen, 5 dozen
>
>   **STEP 3** Decide what arithmetic operation to use.
>
>   *multiplication key word:* per
>
>   **STEP 4** 14 cents per dozen × 5 = **70 cents**

$$\begin{array}{r} 14 \\ \times\ 5 \\ \hline 70\ cents \end{array}$$

In the problems below, underline the necessary information, and circle the multiplication key words. Then solve the problem.

1. Artificially flavored vanilla ice cream costs 92 cents a pint. All-natural vanilla ice cream costs twice as much. How much does the all-natural ice cream cost?

2. Alan needs to buy 4 sets of guitar strings. There are 6 strings per set. How many strings will he buy?

3. Honest Furniture Company's advertisement plays on the radio five times a day and appears in twelve newspapers. How many times does its ad play on the radio in a week?

4. The We Fix-it Company charges $75 per hour to repair computers. The We Fit-it repairman worked for 3 hours repairing computers at the Long Distance Trucking Company. How much did the Long Distance Trucking Company pay for this work?

Identifying Division Key Words

As you should suspect by now, there are also division key words.

EXAMPLE 1 Ron and Nancy shared equally the cost of a $86 phone bill. How much did each of them pay?

division key words: shared equally, each

> **Remember:** Any word indicating that something is cut up is a division key word.

EXAMPLE 2 The 9 million dollar lottery prize will be divided equally among the 3 winners. How much money will each winner receive?

division key words: divided equally, each

Each is considered a division key word, since it indicates that you are given many things and are looking for one.

Circle the division key words. DO NOT SOLVE!

1. Carlos, Dan, and Juan share the driving equally when they drove from Chicago to Los Angeles. How much did Carlos drive according to the map below?

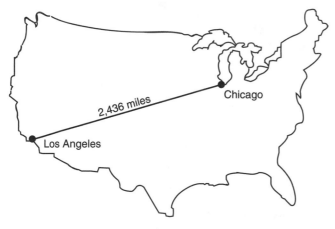

2. A bakery produced 6,300 chocolate chip cookies in a day. The cookies were packed in boxes with 36 cookies per box. How many boxes were used that day?

3. Three salesmen sold $2,250 worth of power tools. On the average, how much did each of them sell?

4. It cost $96 to rent the gym for the basketball game. If the 12 players shared the cost equally, how much did each of them pay?

Solving Division Word Problems with Key Words

Knowing the division key words can help you solve division word problems.

EXAMPLE 1 Union dues of $104 a year can be divided into 52 weekly payments. How much is a weekly payment?

STEP 1 *question:* How much is a weekly payment?

STEP 2 *necessary information:* $104 a year, 52 weekly payments

STEP 3 *division key words:* divided

STEP 4 $104 ÷ 52 weekly payments = **$2**

$$52\overline{)104}^{2}$$

EXAMPLE 2 A 48-minute basketball game is divided into four equal periods. How long is each period?

STEP 1 *question:* How long is each period?

STEP 2 *necessary information:* 48 minutes, 4 periods

STEP 3 *division key words:* divided, equal, each

STEP 4 48 minutes ÷ 4 periods = **12 minutes**

$$4\overline{)48}^{12}$$

For the problems below, underline the necessary information, and circle the division key words. Then solve the word problem.

1. The chocolate bar at the right was shared equally among four children. How much chocolate did each child receive?

12 oz

CHOCO

2. A 60-minute hockey game is divided into three equal periods. How long is the third period?

3. A washing machine that costs $319 when new costs $156 used and can be paid for in 12 monthly payments. How much is each payment on the used washer?

4. A package of 24 mints costs 96 cents. How much does each mint cost?

5. Raffle tickets cost $3 each. If the prizes are worth $4,629, how many tickets must be sold for the raffle to break even?

Key Word Lists for Multiplication and Division

As with addition and subtraction, you can compile lists of multiplication and division key words.

Generally, in multiplication word problems, you are given one of something and asked to find many. You can also think of these problems as multiplying together two parts to get a total.

MULTIPLICATION KEY WORDS	
multiplied	as much
times	twice
total	by
of	area
per	volume

Generally, in division word problems, you are given many things and asked to find one. You can also think of these problems as dividing a total by a part to get the other part.

DIVISION KEY WORDS	
divided (equally)	average
split	every
each	out of
cut	ratio
equal pieces	shared

Remember: Key words are only a clue for solving a problem. Any key word can also appear in word problems needing the opposite operation in order to be solved. You need to read for understanding.

Mental Math with Multiplication and Division

An important mental math tool is multiplying or dividing by 10, 100, or 1,000. Try to multiply 15 by 10. What is your result?

How about 15×100? $15 \times 1,000$?

$15 \times 10 = 150$. $15 \times 100 = 1,500$. $15 \times 1,000 = 15,000$. Do you see a pattern? Test the pattern with other numbers.

Mental Math: When you multiply by 10, 100, or 1,000, mentally add one, two, or three zeros to the right of the number being multiplied. You can also think of this as moving the decimal point to the right one, two, or three places.

Now try dividing 240 by 10. What is your result?

How about $240 \div 100$? $240 \div 1,000$? What is the pattern? Test this pattern with other numbers.

Mental Math: When you divide by 10, 100, or 1,000, move the decimal point of the number being divided to the left one, two, or three places. If the number ends with zeros, you can also mentally subtract one, two, or three zeros.

When you multiply or divide by 10, 100, or 1,000 throughout this book or in your daily life, use mental math either to add on or take off the correct number of zeros or to move the decimal point the correct number of places.

 Mentally multiply these problems.

1. $55 \times 100 =$ $72.6 \times 1,000 =$ $450 \times 10 =$

2. $43 \times 1,000 =$ $99 \times 100 =$ $700 \times 10 =$

 Mentally divide these problems.

3. $38,000 \div 100 =$ $4,493 \div 10 =$ $772 \div 1,000 =$

4. $480 \div 10 =$ $2,972 \div 100 =$ $1,092 \div 1,000 =$

Solving Multiplication and Division Problems with Key Words

In this exercise, some of the word problems have multiplication key words and some have division key words. In each problem, circle the key words. Identify the circled word as a multiplication or division key word. Then solve the problem.

1. The pizza shown at the right is to be divided equally among four people. How many pieces will each person get?

2. An oil well made a profit of $90,000 last year. How much money will each of the five investors receive if all of the profits are divided equally among them?

3. A gas station owner charges $22 per oil change. In one day he did 15 oil changes. What was the total amount of money he received for oil changes?

4. In the discount store, a dress cost $47. In an expensive downtown store, the same dress cost twice as much. How much did the dress cost at the expensive store?

5. To earn a high school equivalency certificate, a student in Illinois must score 225 points on five tests but no less than 40 points on each test. What is the average score on each test that a student needs to get the certificate?

6. Marge's new car averages 28 miles per gallon of gas. How many miles can she expect to drive after she has filled her 15-gallon tank?

Deciding When to Multiply and When to Divide

Word problems are rarely so simple that you can solve them just by finding key words. You must develop your comprehension of the meaning of word problems. Key words are an aid to that understanding.

In earlier chapters, you learned that the same key word that helped you decide to add in one problem might also appear in a subtraction problem. The same is true with multiplication and division key words.

But Don't Despair!

Learning what the key words mean is the first step to understanding word problems.

In the examples that follow, you are given two numbers and are asked to find a third. In each example, you must decide whether to multiply or divide.

The question will ask you to find a total amount, or it will give you a total amount and ask you to find a part.

- When you are given the parts and asked to find the total, you should multiply.

- When you are given the total and a part and you are asked to find a missing part, you should divide.

To get a better idea of this and of what is meant by *part* and *total,* read the two examples on the next page.

EXAMPLE 1 Each of the city's 24 snowplows can plow 94 miles of road a day. If all snowplows are running, how many miles of road can be plowed by the city plows in one day?

STEP 1 *question:* How many miles of road can be plowed?

STEP 2 *necessary information:* 24 snowplows, 94 miles of road

STEP 3 Decide what arithmetic operation to use. Use the following method to draw a diagram.

Draw two boxes, and label them *part* and *part*.

Put another box above them, and label it *total*.

Fill in the boxes with information from the problem. Use only the information that is needed to solve the problem.

Put 24 in the first *part* box, since that is the number of snowplows. Put 94 in the other *part* box, since that is the number of miles each snowplow can plow.

The box that is empty represents the amount that you are looking for, the total number of miles of road.

Since the box representing the total is empty, you should multiply.

STEP 4 snowplows × miles per snowplow = total miles

24 snowplows × 94 miles = **2,256 miles**

$$
\begin{array}{r}
94 \\
\times\ 24 \\
\hline
376 \\
1\ 88 \\
\hline
2{,}256
\end{array}
$$

EXAMPLE 2 A city has 24 snowplows to plow its 2,256 miles of road. How many miles of road must each snowplow cover in order to plow all the city's roads?

STEP 1 *question:* How many miles of road must each snowplow cover?

STEP 2 *necessary information:* 24 snowplows, 2,256 miles of road

STEP 3 Draw and label the boxes.

Put 24 in the first *part* box, since that is the number of snowplows. Put 2,256 in the *total* box, since that has been given as the total number of miles.

The box that is empty represents the amount that you are looking for.

Since the total has been given, you should divide to find the missing part.

STEP 4 total miles ÷ snowplows = miles per snowplow

2,256 miles ÷ 24 snowplows = **94 miles**

$$
\begin{array}{r}
94 \\
24\overline{)2,256} \\
2\,16 \\
\hline
96 \\
96 \\
\hline
0
\end{array}
$$

Examples 1 and 2 are really discussing the same situation. In the first example, the number of plows and miles for each plow are given; you are asked to find the total number of miles that can be covered, and you should multiply. In the second example, the total number of miles are given as well as the number of plows to be used. In this case, you are looking for a missing part (the miles for each plow) and should divide.

....

For each problem, draw *part* and *total* boxes. Then solve the problem.

1. A supermarket sold 78 cartons of Dixie cups. There were 50 cups in every carton. How many cups were sold?

2. Juanita spends an average of $16 a day for food for her family. How much did she spend during the 30-day month of June?

 3. The *Washington Post*'s morning edition was 140 pages long, and the evening edition was 132 pages long. Of each edition, 780,000 copies were printed. How many pages of newsprint were needed to print the morning edition?

4. Fernando's car gets 18 miles per gallon. How many miles can he drive on 21 gallons of gasoline?

 5. In one year, 46,720 people died in car accidents. What was the average number of deaths each day?

6. A factory produces 68,400 nails a day. Every box is packed with 150 nails. How many boxes does the factory need in one day?

Using Diagrams When Deciding to Multiply or Divide

Drawing a picture or diagram is one important strategy for deciding whether to multiply or divide in order to solve a word problem. Look at how a picture or diagram could have helped you solve Example 1 from page 70.

EXAMPLE 1 Each of the city's 24 snowplows can plow 94 miles of road a day. If all snowplows are running, how many miles of road can be plowed by the city plows in one day?

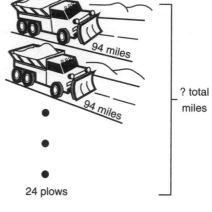

STEP 1 *question:* How many miles of road can be plowed?

STEP 2 *necessary information:* 24 snowplows, 94 miles of road

STEP 3 Draw a diagram and decide what arithmetic operation to use.

The diagram shows each of the 24 snowplows plowing 94 miles of road. To find the total miles, you should multiply.

STEP 4 Do the arithmetic.

24 × 94 = **2,256 total miles**

$$\begin{array}{r} 94 \\ \times\ 24 \\ \hline 376 \\ 188 \\ \hline 2,256 \end{array}$$

STEP 5 Make sure the answer is sensible by using estimation.

Over 20 snowplows must each plow nearly 100 miles. An answer near 2,000 miles makes sense.

EXAMPLE 2 A pint of floor wax covers 2,400 square feet of floor. How many pints of floor wax are needed to wax the 168,000-square-foot floor of the airline terminal?

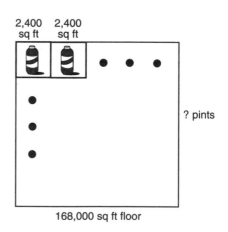

STEP 1 *question:* How many pints of floor wax are needed?

STEP 2 *necessary information:* 1 pint per 2,400 square feet, 168,000 square feet

STEP 3 Draw a diagram and decide what arithmetic operation to use.

The diagram shows that the 168,000-square-foot airline terminal floor must be divided into sections, each 2,400 square feet (the amount covered by one pint of floor wax). To find the number of pints of floor wax, you should divide.

STEP 4 Do the arithmetic.

168,000 ÷ 2,400 = **70 pints of wax**

$$
\begin{array}{r}
70 \\
2{,}400\overline{)168{,}000} \\
\underline{168\,00} \\
00 \\
\underline{00}
\end{array}
$$

STEP 5 Make sure the answer is sensible.

You must wax almost 200,000 square feet divided into 2,500 square feet sections. 200,000 ÷ 2,500 = about 80, so 70 pints makes sense.

Each word problem is followed by two diagrams with short explanations. One choice of a diagram and explanation gives you a reason to multiply to find the answer. The other gives you a reason to divide to find the answer. Circle the letter of the correct explanation and solve the problem.

1. Larry feeds each of his 380 laboratory animals 5 ounces of food pellets a day. How many ounces of food pellets does he need for one day?

   a. Since each animal eats 5 ounces, multiply to find the total ounces needed.

380 lab animals

5 oz of food

? oz total food needed

   b. To find the amount of food available, divide the total animals (380) by the amount of food available for each.

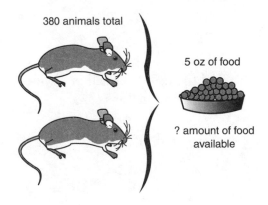

380 animals total

5 oz of food

? amount of food available

2. Tickets to the play were $6. At the end of the night, Janet counted $3,102 in receipts for the performance. How many people bought tickets for the play?

    a. To find the total $ for people, multiply the cost of a ticket ($6) by the receipts ($3,102).

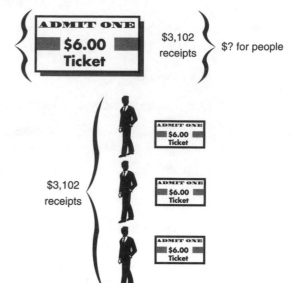

    b. To find the number of people, divide the amount of receipts ($3,102) by the price of one ticket ($6).

3. After filling up her gas tank, Jean drove 260 miles. After the drive, she refilled her tank with 13 gallons of gas. On the average, how many miles was she able to drive on a gallon of gas?

    a. To find the total miles, multiply the miles for a car (260) by the gallons (13).

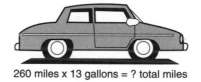

260 miles x 13 gallons = ? total miles

    b. To find the miles per gallon, divide the miles (260) by the number of gallons (13).

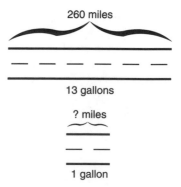

4. Glennie planted 8 rows of tomatoes in her truck garden. If she planted 48 plants in each row, how many tomato plants did she plant?

   a. Since there are 8 rows and 48 plants in each row, you should multiply.

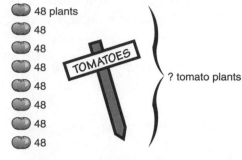

   b. To find the number of plants, divide 48 plants by 8 rows.

5. Shirley, a buyer for a major department store, has a budget of $6,000 to buy 400 blouses. What is the most she can pay per blouse?

   a. To find the most Shirley can pay for one blouse, divide the total budget ($6,000) by the number of blouses (400).

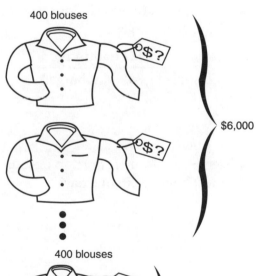

   b. To find the total paid, multiply the budget ($6,000) by the number of blouses (400).

Underline the correct phrase by using the solution given after each word problem. The first one is done for you.

6. A football television contract for $78,000,000 is to be _____ 60 colleges. How much will each college receive?

$$\overset{\$\ 1,300,000}{60\ \text{colleges}\,\big)\,\$78,000,000}$$ (<u>divided equally among</u>, given to each of)

7. Mary bought 4 skirts for $24 _____. How much did she spend?

$24
× 4 skirts   (total, each)
$96

8. There were 36,000 trees _____ in the state forest before the fire. During the fire, 48 square miles of forest burned. How many trees were destroyed in all?

36,000 trees
×   48 square miles   (total, per square mile)
1,728,000 trees

9. During an average 12-hour workday, the fast-food restaurant sold 3,852 hamburgers.

$$\overset{\textbf{321 hamburgers}}{12\ \text{hours}\,\big)\,3,852\ \text{hamburgers}}$$ (How hamburgers were sold in a week?, How many hamburgers were sold per hour?)

10. A cafeteria serves 3,820 _____ a day, with each person being served an 8-ounce portion of soup. How many ounces of soup must be made in one day?

3,820
×   8 oz of soup   (ounces of soup, people)
30,560 oz of soup

MULTIPLICATION AND DIVISION WORD PROBLEMS: DECIMALS AND FRACTIONS

Solving Decimal Multiplication and Division Word Problems

Fraction and decimal word problems are solved in the same ways as word problems using whole numbers.

The following examples show a method for solving multiplication and division word problems containing decimals. As with whole-number word problems, multiply when you are looking for the total, and divide when you are looking for one of the parts. Estimation can be very helpful with these problems.

EXAMPLE 1   Gasoline costs $1.499 per gallon. How much do 18 gallons of gasoline cost?

STEP 1   *question:* How much do 18 gallons of gasoline cost?

STEP 2   *necessary information:* $1.499 per gallon, 18 gallons

STEP 3   *diagram:*

```
        ┌───────┐
        │   ?   │
        └───────┘
          total

      ┌──────┐┌───────┐
      │  18  ││ 1.499 │
      └──────┘└───────┘
       part     part
```

STEP 4   price of each gallon × number of gallons = total cost

18 × $1.499 = 26.982 = **$26.98**

(In money problems that have answers containing more than two decimal places, you should round your answer to the nearest cent.)

$$\begin{array}{r} 1.499 \\ \times\quad 18 \\ \hline 11992 \\ 1499 \\ \hline 26.982 \end{array}$$

STEP 5   *estimation:* 18 × 1.5 = 27

This estimation shows that the answer is sensible.

EXAMPLE 2 Gil's car gets 28.6 miles per gallon. Last month he drove 943.8 miles. How many gallons of gas did he need for the month?

> **STEP 1** *question:* How many gallons of gas did he need for the month?
>
> **STEP 2** *necessary information:* 28.6 miles per gallon, 943.8 miles
>
> **STEP 3** *diagram:*

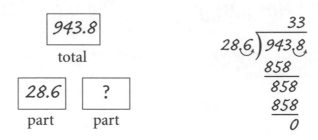

> **STEP 4** total miles ÷ miles per gallon = gallons
>
> 943.8 miles ÷ 28.6 miles per gallon = **33 gallons**
>
> **STEP 5** *estimation:* 900 ÷ 30 = 30

Calculator: Your calculator can be very useful when multiplying or dividing decimals. But you must be careful about putting the decimal in the correct place. It is important that you estimate the answer to your problem before doing the calculator work, so you will know that you entered the numbers correctly on the calculator.

   If you made a mistake entering the decimal point and got a result of 3.3 gallons, your estimate should alert you.

For each problem, use *part* and *total* boxes to help you decide whether to multiply or divide. Then solve the problem. Remember to write the label of the answer and to round all money problems to the nearest cent.

1. A runner ran an average of 6.5 minutes per mile for a race that had 242 official entrants. If the race was 6.2 miles long, how long did it take him or her to run it?

    total

    part   part

2. A nonprofit food co-op bought a 40-pound sack of onions for $23.20. How much will the co-op members pay per pound if the onions are sold at cost?

    total

    part   part

3. After filling her gas tank, Katie drove 159.75 miles. After the ride, she filled her tank again with 7.1 gallons of gas. On the average, how many miles per gallon did she get on the trip?

total

part   part

4. A beef round roast costs $2.29 per pound. How much is a 4.67-pound roast?

total

part   part

5. Four roommates share their food bill equally. Last month they spent $372.36 for food and $850.00 for rent. How much did each of them pay for food?

total

part   part

6. A salesman, working on commission, earned $118.56 in an 8-hour workday. On the average, how much did he earn each hour?

total

part   part

7. It costs $0.85 to ride the city bus. At the end of the day, Veronique emptied the bus's cash box and deposited $167.45 in fares. How many passengers rode the bus that day?

total

part   part

8. Lynelle spent $3.65 for transportation every work day. Last year she worked 239 days. How much did she spend on transportation during work days last year?

total

part   part

Solving Fraction Multiplication Word Problems

When you multiply two whole numbers, the answer is larger than either number. But when you multiply a number by a fraction smaller than 1, the answer is smaller than the original number. For example, $21 \times \frac{2}{3} = 14$.

Multiplication and division word problems with fractions often seem confusing. When you multiply by a fraction, you may end up with a smaller number, and when you divide by a fraction, you may end up with a larger number. This is the opposite of what you have come to expect with whole numbers.

The following chart should help you remember when to expect a larger or smaller answer when multiplying or dividing.

When Multiplying a Number By:	Your Answer Will Be:	Example:
a number greater than 1	larger than the number	$36 \times 2 = 72$
1	the same as the number	$36 \times 1 = 36$
a fraction smaller than 1	smaller than the number	$\overset{9}{36} \times \frac{3}{\underset{1}{4}} = 27$
(**Remember:** *An improper fraction is greater than 1. For example,* $36 \times \frac{4}{3} = 48$.)		

When Dividing a Number By:	Your Answer Will Be:	Example:
a number greater than 1	smaller than the number	$36 \div 2 = \overset{18}{36} \times \frac{1}{\underset{1}{2}} = 18$
1	the same as the number	$36 \div 1 = 36$
a fraction smaller than 1	larger than the number	$36 \div \frac{3}{4} = \overset{12}{36} \times \frac{4}{\underset{1}{3}} = 48$
(*Dividing by an improper fraction is the same as multiplying by a fraction less than 1. For example,* $36 \div \frac{4}{3} = \overset{9}{36} \times \frac{3}{\underset{1}{4}} = 27$.)		

The most common key word in fraction multiplication problems is *of*—as in *finding a fraction of* something. Some people confuse these problems with division because they require you to find a piece of something. The example below illustrates why you multiply when you find a fraction of a quantity.

Find $\frac{1}{2}$ of 6.

You already know that this is 3. Multiplying by $\frac{1}{2}$ gives the same result as dividing by 2. When you multiply the two numbers, you really multiply the numerators (the numbers above the line) and divide by the denominator (the number below the line).

$$\frac{1}{2} \times 6 = \frac{1}{1}\frac{1}{2} \times \frac{6^3}{1} = \frac{3}{1} = 3$$

The following examples show you how to solve multiplication word problems that require you to find a fraction of a quantity.

EXAMPLE 1 Bernie's Service Station inspected 20 cars yesterday. Of the 20 cars inspected, $\frac{1}{5}$ failed the inspection. How many cars failed the inspection?

> **STEP 1** *question:* How many cars failed the inspection?
>
> **STEP 2** *necessary information:* 20 cars, $\frac{1}{5}$ of the cars
>
> **STEP 3** *key word:* of
>
> fraction (of) × total = part
>
> **STEP 4** $\frac{1}{5}$ × 20 cars = cars that failed
>
> $\frac{1}{1\cancel{5}} \times \frac{\cancel{20}^4}{1} =$ **4 cars**

Some multiplication word problems that involve fractions do not have the key word *of*. These problems can be recognized as multiplication, since you are usually given the size of one item and asked to find the size of many. To go from one to many, you should multiply.

EXAMPLE 2 In a high school, class periods are $\frac{3}{4}$ hour long. How long will 8 periods last?

> **STEP 1** *question:* How long will 8 periods last?
>
> **STEP 2** *necessary information:* $\frac{3}{4}$ hour, 8 periods
>
> **STEP 3** You are given the length of one class period ($\frac{3}{4}$ hour) and are asked to find the total length of many class periods (8 periods). Therefore, you should multiply.
>
> **STEP 4** 8 periods × $\frac{3}{4}$ hour = $^2\cancel{8} \times \frac{3}{\cancel{4}_1} =$ **6 hours**

When you are working with a word problem and have to decide whether to multiply or divide, it is especially helpful to use **Step 5: Check to see that your answer is sensible.**

In Example 2, if you had mistakenly divided 8 by $\frac{3}{4}$, your answer would have been $10\frac{2}{3}$ hours. ($8 \div \frac{3}{4} = 8 \times \frac{4}{3} = 10\frac{2}{3}$.) Would it make sense to say that 8 periods, each consisting of less than 1 hour, would total $10\frac{2}{3}$ hours?

In the following problems, underline the necessary information. Then solve the problem.

1. In Chicago last year $\frac{2}{3}$ of the precipitation was rain. According to the chart, how many inches of rain fell in Chicago?

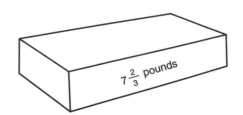

Precipitation	
Baltimore	27 inches
Chicago	36 inches
New York	40 inches

2. Of the car accidents in the state last year, $\frac{7}{8}$ were in urban areas. There were 23,352 car accidents in the state last year. How many accidents were in urban areas?

3. How much do 10 of the boxes shown at the right weigh?

$7\frac{2}{3}$ pounds

4. Brad's dog Cedar eats $\frac{2}{3}$ can of dog food and two dog biscuits a day. How many cans of dog food will Brad need to feed Cedar for 12 days?

5. A candy bar contains $1\frac{1}{8}$ ounces of peanuts. How many ounces of peanuts are in $3\frac{1}{2}$ candy bars?

6. Only $\frac{1}{2}$ cup of a new concentrated liquid detergent is needed to clean a full load of laundry. How much detergent is needed to clean $\frac{1}{2}$ of a load of laundry?

7. In a $\frac{1}{4}$-pound hamburger, $\frac{2}{5}$ of the hamburger meat was fat. How much fat was in the hamburger?

8. A space shuttle was traveling 17,000 miles per hour. How far did it travel in $2\frac{1}{2}$ hours?

Solving Fraction Division Word Problems

Remember that the second number in a fraction division problem will be inverted (turned upside down). Therefore, it is very important that the total amount that is being divided is always the first number that you write when solving such a problem. But even though the total amount comes first when you are solving the fraction division problem, it does not always appear first in a word problem.

EXAMPLE 1 Joyce made dinner for nine people. She divided a $\frac{1}{4}$-pound stick of butter equally among them. How much butter did each person receive?

STEP 1 *question:* How much butter did each person receive?

STEP 2 *necessary information:* $\frac{1}{4}$ pound, 9 people

STEP 3 *key words:* divided equally, each

Nine people are sharing the butter. To find how much butter one person receives, you should divide.

STEP 4 total butter ÷ number of people = butter per person

$\frac{1}{4}$ pound ÷ 9 people = $\frac{1}{4} \times \frac{1}{9} = \frac{1}{36}$ **pound per person**

> **Remember:** The total amount is not always the largest number.

Many division word problems contain the concept of cutting a total into pieces. If you are given the size of the total, you should divide to find a part—either the number of pieces or the size of each piece.

EXAMPLE 2 Quality Butter Company makes butter in 60-pound batches. It then cuts each batch into $\frac{1}{4}$-pound sticks of butter. How many sticks of butter are made from each batch?

STEP 1 *question:* How many sticks of butter are made from each batch?

STEP 2 *necessary information:* 60-lb batches, $\frac{1}{4}$-lb sticks

STEP 3 *key words:* cuts, each

total amount ÷ size of each piece = number of pieces

STEP 4 60-pound batches ÷ $\frac{1}{4}$-pound sticks = $\frac{60}{1} \div \frac{1}{4} = \frac{60}{1} \times \frac{4}{1} = 60 \times 4 = $ **240 sticks**

Underline the necessary information in each problem below. Then solve the problem.

1. A box of $22\frac{1}{2}$ inches deep. How many books can be packed in the box if each book is $\frac{5}{8}$ inch thick?

2. Gloria is serving a dinner for 13 people. She is cooking a $6\frac{1}{2}$-pound roast. How much meat would each person get if she divided the roast equally?

3. A bookstore gift wraps books using $2\frac{1}{4}$ feet of ribbon for each book. How many books can the store gift wrap from a roll of ribbon $265\frac{1}{2}$ feet long?

4. A container contains $8\frac{1}{2}$ pounds of mashed potatoes. Linh, who works in a cafeteria, divide the potatoes into servings the size shown at the right. How many servings can she make from the container of potatoes?

Mashed Potatoes

$\frac{1}{2}$ pound

5. A can of Diet Delight peaches contains $9\frac{3}{4}$ ounces of peaches. If one can is used for three equal servings, how large would each serving be?

6. Tatiana wants to divide the garden shown at the right into $1\frac{1}{2}$-foot-wide sections. How many sections can she make?

12 feet

Solving Fraction Multiplication and Division Word Problems

Underline the necessary information and decide whether to multiply or divide. Then solve the problem.

1. A music practice room is used 12 hours a day. If each practice session is $\frac{3}{4}$ hour long, how many sessions are there in a day?

2. At full production, a car rolls off the assembly line every $\frac{2}{3}$ hour. At this rate, how long does it take to produce 30 cars?

3. At full production, a car rolls off the assembly line every $\frac{2}{3}$ hour. At this rate, how many cars are produced in 24 hours?

4. A consumer group claimed that $\frac{2}{3}$ of all microwave ovens were defective. The chart shows oven sales in one state last year. According to the consumer group's findings, how many microwave ovens sold in this state would have been defective?

Oven Sales	
Microwave	26,148
Regular	59,882

5. On a wilderness hike, six hikers had to share $4\frac{1}{2}$ pounds of chocolate bar and $1\frac{3}{4}$ pounds of dry milk. If it was cut equally, how much chocolate did each hiker receive?

6. A recipe calls for the ingredients at the right. If a cook wants to double the recipe, how much baking soda will he or she need?

1 tsp baking powder

$1\frac{1}{2}$ tsp baking soda

$\frac{1}{2}$ tsp salt

Put a check next to the correct question to complete each problem below. Use the solution given after each word problem.

7. For her printer, Awilda bought a toner cartridge that contained 3.5 ounces of toner. The toner is supposed to last for 5,000 copies.
   (_____ On the average, how many copies can be made per ounce of toner?)
   (_____ On the average, how much toner is used for each copy?)

$$\begin{array}{r} 0.0007 \\ 5{,}000\overline{)3.5000} \end{array}$$

8. Elena used to pay $0.13 a minute to call her mother long distance. Since she changed to her new phone company, she now pays $0.09 a minute. This month she talked to her mother for 56 minutes.
   (_____ What will the new phone company charge for her talks with her mother?)
   (_____ What would her old phone company charge for her talks with her mother?)

$$\begin{array}{r} \$0.09 \\ \times\ \ 56 \\ \hline \$5.04 \end{array}$$

9. In order to cut down the amount of fat in her family's diet, Alejandrina decided to use only $\frac{2}{3}$ of the butter called for in a recipe. Her recipe for brownies originally called for $\frac{1}{2}$ cup of butter.
   (_____ How much butter should she use if she wanted to double the brownie recipe?)
   (_____ How much butter should she use for her new brownie recipe?)

$\frac{1}{2} \times \frac{2}{3} = \frac{1}{3}$ cup

10. A package of 30 fig bars weighs 16 ounces.
    (_____ What is the total weight of 16 packages of fig bars?)
    (_____ How much does a single fig bar weigh?)

$$\begin{array}{r} 0.53 \quad \text{or } 5\frac{1}{3} \text{ oz} \\ 30\overline{)16.0000} \\ \underline{150} \\ 100 \\ \underline{90} \\ 100 \end{array}$$

Solving Multiplication and Division Word Problems

For each problem, circle the letter of the correct answer. Round money problems to the nearest cent and other decimal problems to the nearest hundredth.

1. Forty pounds of mayonnaise were packed in jars that weighed $\frac{1}{8}$ pound and could hold $\frac{5}{8}$ pound of mayonnaise. How many jars were needed to pack all the mayonnaise?

   a. 25 jars
   b. 64 jars
   c. 41 jars
   d. 320 jars
   e. 5 jars

2. A 32-square-foot piece of $\frac{1}{4}$-inch-thick plywood costs $20.80. How much does it cost per square foot?

   a. $52.80
   b. $11.20
   c. $0.65
   d. $12.80
   e. $5.20

3. To cover the cost of the prizes, a VFW Post had to sell at least $\frac{1}{6}$ of the raffle tickets. They had 3,000 raffle tickets printed. How many tickets did they have to sell?

   a. 5,000 raffle tickets
   b. 500 raffle tickets
   c. 1,800 raffle tickets
   d. 18,000 raffle tickets
   e. none of the above

4. The trainer of the championship baseball team was voted $\frac{3}{5}$ of a winner's share. If a winner's share is $17,490 and there were 40 shares, how much money did the trainer receive?

   a. $29,155
   b. $10,494
   c. $437.25
   d. $3,498
   e. $728.75

5. To make one apron, Janice needed $\frac{2}{3}$ yard of cloth. She has a roll of cloth $7\frac{1}{3}$ yards long. If she doesn't waste any cloth, how many aprons can she make by cutting and using the entire roll of cloth?

    a. $4\frac{8}{9}$ aprons
    b. 4 aprons
    c. 5 aprons
    d. 11 aprons
    e. 8 aprons

6. Max used 8.1 gallons of gas when he drove 263.1 miles in 4.5 hours. What was his average speed for the trip? (Round to the nearest tenth.)

    a. 31.1 miles per hour
    b. 58.3 miles per hour
    c. 58.4 miles per hour
    d. 58.5 miles per hour
    e. 31.2 miles per hour

7. The Motown Music Company shipped 1,410 CDs to the Midtown Music Store. If 30 CDs were packed in each box, how many boxes were needed to ship the CDs?

    a. 423 boxes
    b. 470 boxes
    c. 47 boxes
    d. 43 boxes
    e. none of the above

8. Ingrid ran in a race from Templeton to Redfield. One kilometer is equal to 0.62 mile. How many miles did she run?

    a. 24 miles
    b. 9.3 miles
    c. 24.19 miles
    d. 4.13 miles
    e. none of the above

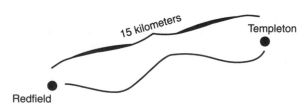

USING PROPORTIONS

What Are Ratios?

A **ratio** is a comparison of two groups. Ratios can be written in a number of ways.

EXAMPLE   A small luncheonette has 8 chairs for 2 tables. The ratio of chairs to tables can be written three ways.

8 chairs for 2 tables

8 to 2, more commonly written as 8:2

$\dfrac{8 \text{ chairs}}{2 \text{ tables}}$

In the rest of this book, you will use only the third way of writing a ratio, the fraction form.

Note: Always write labels for both the top and the bottom of the ratio.

Write the following relationships as ratios in the fraction form. The first problem has been done for you.

1. 1 customer bought 6 cans of tomato soup. $\dfrac{1 \text{ customer}}{6 \text{ cans}}$

2. 2 teachers worked with 30 students.

3. Phi Hung earned 120 dollars in 8 hours.

4. Yvette drove 38 miles on 2 gallons of gasoline.

5. The company provided 3 buses for 114 commuters.

What Are Proportions?

A **proportion** expresses two ratios that have the same value. In arithmetic, you have studied these as equivalent fractions. For example, $\frac{75}{100} = \frac{3}{4}$.

EXAMPLE 1 The center of the city has 1 bus stop every 3 blocks. Therefore, the city center has 2 bus stops every 6 blocks.

$$\frac{1 \text{ bus stop}}{3 \text{ blocks}} = \frac{2 \text{ bus stops}}{6 \text{ blocks}}$$

EXAMPLE 2 The ratio of women to men working at the Small Motors Repair Shop is 3 women to 4 men. If there are 8 men working at the repair shop, how many women work there?

One of the numbers of the proportion is not given: the number of women working at the repair shop. Therefore, when the proportion is written, a placeholder is needed in the place where the number of women should be written. The letter *n*, standing for a number, is used as the placeholder, but any letter could be used.

$$\frac{3 \text{ women}}{4 \text{ men}} = \frac{n \text{ women}}{8 \text{ men}}$$

Finding the number that belongs in place of the *n* is called **solving a proportion.** Look at two methods that can be used to solve a proportion.

Method 1: Multiplication

STEP 1 *question:* How many women work there?

STEP 2 *necessary information:* 3 women, 4 men, 8 men

STEP 3 Write a proportion based on the problem.

$$\frac{3 \text{ women}}{4 \text{ men}} = \frac{n \text{ women}}{8 \text{ men}}$$

$$\frac{3 \times \Box}{4 \times \boxed{2}} = \frac{n}{8}$$

Notice that both denominators have been filled in and that $4 \times 2 = 8$.

$$\frac{3 \times \boxed{2}}{4 \times \boxed{2}} = \frac{6}{8}$$

Since proportions are equivalent fractions, you multiply the top and the bottom by the same number. In this problem, the number is 2.

STEP 4 Therefore, $3 \times 2 = $ **6 women.**

There are cases when Method 1 does not work as simply. This is especially true when a problem contains a decimal or a fraction, or when the numbers are not simple multiples of each other. In these problems, Method 2 is quite useful.

Method 2: Cross Multiplication

STEP 1 *question:* How many women work there?

STEP 2 *necessary information:* 3 women, 4 men, 8 men

STEP 3 Write the proportion.

$$\frac{3 \text{ women}}{4 \text{ men}} = \frac{n \text{ women}}{8 \text{ men}}$$

$$\frac{3}{4} \diagup\!\!\!\!\diagdown \frac{n}{8}$$

STEP 4 Cross multiply. Multiply the numbers that are on a diagonal.

$$4 \times n = 3 \times 8$$
$$4n = 24$$

> **Note:** The letter is usually written on the left side. Also, $4n$ means the same as 4 times n. It is not necessary to write the multiplication sign, $\times$.

STEP 5 To find n, the number of women, divide the number standing alone by the number next to the letter.

$$n = \frac{24}{4} = 6$$

$n =$ **6 women**

Notice that when you write a proportion, the labels must be consistent. For example, if *women* is the label of the top of one side of a proportion, it must be on the top of the other side.

<u>EXAMPLE 3</u> Chin has seen 6 movies in the last 9 months. At this rate, how many movies will she see in 12 months?

STEP 1 *question:* How many movies will she see in 12 months?

STEP 2 *necessary information:* 6 movies, 9 months, 12 months

STEP 3 Write the proportion.

$$\frac{12 \text{ months}}{n \text{ movies}} = \frac{9 \text{ months}}{6 \text{ movies}}$$

$$\frac{12}{n} \diagup\!\!\!\!\diagdown \frac{9}{6}$$

STEP 4 Cross multiply.

$$9 \times n = 12 \times 6$$
$$9n = 72$$

STEP 5 Divide.

$$n = \frac{72}{9} = 8$$

$n =$ **8 movies**

<u>EXAMPLE 4</u> Sandy read that she should cook a roast 20 minutes for each half pound. How large a roast could she cook in 90 minutes?

STEP 1 *question:* How large a roast could she cook in 90 minutes?

STEP 2 *necessary information:* $\frac{1}{2}$ pound, 20 minutes, 90 minutes

STEP 3 Write the proportion.

$$\frac{\frac{1}{2} \text{ pound}}{20 \text{ minutes}} = \frac{n \text{ pounds}}{90 \text{ minutes}}$$

$$\frac{\frac{1}{2}}{20} \times \frac{n}{90}$$

STEP 4 Cross multiply.

$$20 \times n = 90 \times \frac{1}{2}$$

STEP 5 Divide.

$$20n = 45$$

$$n = 2\frac{1}{4} \textbf{ pounds of roast}$$

$$n = \frac{45}{20} = 2\frac{1}{4}$$

···

Solve the following proportions for *n*.

1. $\dfrac{160 \text{ miles}}{5 \text{ hours}} = \dfrac{n \text{ miles}}{10 \text{ hours}}$

2. $\dfrac{12 \text{ cars}}{32 \text{ people}} = \dfrac{3 \text{ cars}}{n \text{ people}}$

3. $\dfrac{n \text{ dollars}}{8 \text{ quarters}} = \dfrac{6 \text{ dollars}}{24 \text{ quarters}}$

4. $\dfrac{42 \text{ pounds}}{n \text{ chickens}} = \dfrac{14 \text{ pounds}}{4 \text{ chickens}}$

5. $\dfrac{28,928 \text{ people}}{8 \text{ doctors}} = \dfrac{n \text{ people}}{1 \text{ doctor}}$

6. $\dfrac{\$24.39}{1 \text{ shirt}} = \dfrac{n \text{ dollars}}{6 \text{ shirts}}$

7. $\dfrac{\$47.85}{3 \text{ shirts}} = \dfrac{n \text{ dollars}}{10 \text{ shirts}}$

8. $\dfrac{3 \text{ minutes}}{\frac{1}{2} \text{ mile}} = \dfrac{n \text{ minutes}}{5 \text{ miles}}$

9. $\dfrac{575 \text{ passengers}}{n \text{ days}} = \dfrac{1,725 \text{ passengers}}{21 \text{ days}}$

10. $\dfrac{7 \text{ blinks}}{\frac{1}{10} \text{ minute}} = \dfrac{n \text{ blinks}}{10 \text{ minutes}}$

Using Proportions to Solve Word Problems

Proportions can be used when you are unsure of whether to multiply or divide.

The following examples show how to write proportions to solve multiplication and division word problems.

EXAMPLE 1 There are 16 cups in a gallon. At the church picnic, Carmella poured 5 gallons of cola into paper cups that each held 1 cup of soda. How many cups did she fill?

STEP 1 *question:* How many cups did she fill?

STEP 2 *necessary information:* 16 cups in a gallon, 5 gallons, 1 cup

STEP 3 Write the proportion.

labels for proportion: $\dfrac{\text{cups}}{\text{gallons}}$

$$\dfrac{16 \text{ cups}}{1 \text{ gallon}} = \dfrac{n \text{ cups}}{5 \text{ gallons}}$$

$$\dfrac{16}{1} \diagup\!\!\!\!\diagdown \dfrac{n}{5}$$

STEP 4 Cross multiply.

(*n* means the same as $1 \times n$. From now on you don't need to write the 1 and the $\times$ sign, so you write $n = 16 \times 5$.)

$n =$ **80 cups**

$n = 16 \times 5$
$n = 80$

Remember: When you write the labels for a proportion, it doesn't matter which category goes on top. But once you make a decision, you must stick with it. Once you put *cups* on the top of one ratio, you must keep *cups* on top of the other.

16 cups in a gallon means the same as $\dfrac{16 \text{ cups}}{1 \text{ gallon}}$. The 1 will often not appear in these word problems. When writing a proportion, you must determine when a 1 is needed and where it goes. You can do this by first identifying the two labels and then putting numbers in the proportion.

There are a number of word phrases that require that a 1 be used in a ratio.

Phrases	Meaning
27 miles per gallon	$\dfrac{27 \text{ miles}}{1 \text{ gallon}}$
$8 an hour	$\dfrac{\$8}{1 \text{ hour}}$
3 meals a day	$\dfrac{3 \text{ meals}}{1 \text{ day}}$
30 miles each day	$\dfrac{30 \text{ miles}}{1 \text{ day}}$

EXAMPLE 2 At the Boardwalk Arcade, owner Manuel Santos collects 1,380 quarters every day. There are 4 quarters in a dollar. How many dollars does he collect every day?

STEP 1 *question:* How many dollars does he collect every day?

STEP 2 *necessary information:* 1,380 quarters, 4 quarters in a dollar

STEP 3 *labels for proportion:* $\dfrac{\text{dollars}}{\text{quarters}}$

$$\frac{n \text{ dollars}}{1,380 \text{ quarters}} = \frac{1 \text{ dollar}}{4 \text{ quarter}}$$

$$\frac{n}{1,380} \diagup\!\!\!\!\diagdown \frac{1}{4}$$

STEP 4 Cross multiply.

$$4 \times n = 1,380 \times 1$$
$$4n = 1,380$$

STEP 5 Divide.

$$n = \frac{1,380}{4} = 345$$

$n = $ **345 dollars**

Underline the necessary information. Write proportions for the problems below and solve them.

1. A shipment of vaccine can protect 7,800 people. How many shipments of vaccine are needed to protect 140,400 people living in the Portland area?

2. It costs $340 an hour to run the 1,000-watt power generator. How much does it cost to run the generator for 24 hours?

3. How many ounces of soup (like the can shown at the right) are in a carton containing 28 cans?

4. Jim types 52 words per minute. How many words can he type in 26 minutes?

5. An elementary school nurse used 3,960 Band-Aids last year. There were 180 school days. On the average, how many Band-Aids did he use a day?

6. The company health clinic gave out 5,460 aspirin and 720 antacid tablets last year. How many bottles of aspirin did the clinic use last year if there were 260 aspirin in a bottle?

7. A coal mine produced 126 tons of slag in a week. Trucks removed the slag in 3-ton loads. How many loads were needed to remove all the slag?

8. Cloth is sold by the yard. Edyth bought the piece of cloth shown at the right to make dresses. There are 3 feet in a yard. How many yards of cloth did she buy?

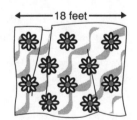

9. A 6-ounce can of water chestnuts contains 26 water chestnuts. Hong used 3 cans of water chestnuts. How many water chestnuts did she use?

Using Proportions to Solve Decimal Word Problems

You can use proportions to solve decimal multiplication and division word problems. The problems should be set up as if the numbers were whole numbers. Multiply or divide as if you were using whole numbers. Then use the rules for decimal multiplication and division to place the decimal point in the right place. Finally, round the answer if necessary.

EXAMPLE 1 Ray has $15.00 to spend on gasoline. How many gallons can he buy if 1 gallon costs $1.20?

STEP 1 *question:* How many gallons can he buy?

STEP 2 *necessary information:* $15.00, $1.20 for a gallon

STEP 3 *labels for proportion:* $\dfrac{\$}{\text{gallons}}$

$$\frac{\$15.00}{n \text{ gallons}} = \frac{\$1.20}{1 \text{ gallon}}$$

STEP 4 Cross multiply.

STEP 5 Divide.

$n =$ **12.5 gallons**

$$\frac{15}{n} \diagdown \frac{1.20}{1}$$

$$1.20 \times n = 15 \times 1$$
$$1.20n = 15$$

$$n = \frac{15}{1.20} = 12.5$$

EXAMPLE 2 There are 236.5 milliliters in a cup. A recipe calls for 3 cups of flour. Maria has only metric spoons and measuring cups. How many milliliters of flour does she need for the recipe?

STEP 1 *question:* How many milliliters of flour does she need for the recipe?

STEP 2 *necessary information:* 236.5 milliliters in a cup, 3 cups

STEP 3 *labels for proportion:* $\dfrac{\text{milliliters}}{\text{cups}}$

$$\frac{236.5 \text{ milliliters}}{1 \text{ cup}} = \frac{n \text{ milliliters}}{3 \text{ cups}}$$

STEP 4 Cross multiply.

$n =$ **709.5 milliliters**

$$\frac{236.5}{1} \diagdown \frac{n}{3}$$

$$n = 3 \times 236.5$$
$$n = 709.5$$

Underline the necessary information. Write the labels for the proportion, fill in the numbers, and solve the proportion. Round your answers to the nearest hundredth.

1. There are 25.4 millimeters in an inch. How many inches long is a 100-millimeter cigarette?

2. A kilogram weight is shown at the right. The police seized 36 kilograms of illegal drugs. How many pounds did the drugs weigh?

1 KILOGRAM
2.2 POUNDS

3. Alba worked 35.5 hours last week. She earns $8.62 an hour. How much money did she earn last week?

4. There are 1.09 yards in a meter. Gary ran in an 880-yard race. How many meters did he run?

5. Cindy spent $20.00 on gasoline. The gasoline cost $1.15 per gallon. How many gallons of gasoline did she buy?

6. There are approximately 1.61 kilometers in a mile. The speedometer on Iris's imported car is in kilometers per hour. She does not want to speed. What is 55 miles per hour in kilometers per hour?

7. A Tiger Milk nutrition bar weighs 35.4 grams. The factory processed 9,486 bars in one run. How many grams of Tiger Milk bars were processed?

Using Proportions to Solve Fraction Word Problems

Proportions can be used to solve multiplication and division word problems containing fractions. Though they look complicated when you set them up, they are manageable after you cross multiply.

EXAMPLE 1 A dump truck can carry a load of $2\frac{3}{4}$ tons of gravel. In one day, the truck removed 8 loads of gravel from a gravel pit. How many tons of gravel did it remove from the pit that day?

STEP 1   *question:* How many tons of gravel did it remove from the pit that day?

STEP 2   *necessary information:* a load of $2\frac{3}{4}$ tons, 8 loads

STEP 3   *labels for proportion:* $\dfrac{\text{tons}}{\text{loads}}$

$$\frac{2\frac{3}{4}\text{ tons}}{1\text{ load}} = \frac{n\text{ tons}}{8\text{ loads}}$$

STEP 4   Cross multiply.

$n = $ **22 tons**

$$\frac{2\frac{3}{4}}{1} \diagup\!\!\!\!\diagdown \frac{n}{8}$$

$$n = 8 \times 2\frac{3}{4}$$

$$n = \frac{\overset{2}{\cancel{8}}}{1} \times \frac{11}{\underset{1}{\cancel{4}}} = 22$$

EXAMPLE 2 Top Burger makes a $\frac{1}{4}$-pound hamburger. How many of these hamburgers can be made from 50 pounds of hamburger meat?

STEP 1   *question:* How many of these hamburgers can be made?

STEP 2   *necessary information:* $\frac{1}{4}$ pound, 50 pounds

STEP 3   *labels for proportion:* $\dfrac{\text{pounds}}{\text{hamburgers}}$

$$\frac{\frac{1}{4}\text{ pound}}{1\text{ hamburger}} = \frac{50\text{ pounds}}{n\text{ hamburgers}}$$

STEP 4   Cross multiply.

STEP 5   Multiply both sides by 4 to clear the *n*.

$n = $ **200 hamburgers**

$$\frac{\frac{1}{4}}{1} \diagup\!\!\!\!\diagdown \frac{50}{n}$$

$$\frac{1}{4} \times n = 50 \times 1$$

$$\frac{1}{4}n = 50$$

$$n = 50 \times 4 = 200$$

A word problem asking you to find a fraction of something is easier to solve by direct multiplication than by using a proportion. Example 3 illustrates this.

EXAMPLE 3 Of all gallons of milk sold in a store, $\frac{2}{5}$ are low fat. The store sold 380 gallons of milk. How many gallons of low-fat milk were sold?

You could set up the following proportion to solve the problem.

$$\frac{\frac{2}{5} \text{ low fat}}{1 \text{ gallon}} = \frac{n \text{ low fat}}{380 \text{ gallons}}$$

While this will give you the correct answer, it is easier to remember that you should multiply to find a fraction of something. It is easier to solve the problem this way.

Fraction (of) × total gallons = low-fat gallons

$$\frac{2}{{}_1\cancel{5}} \times \frac{\cancel{380}^{76}}{1} = \frac{2}{1} \times \frac{76}{1} = \textbf{152 gallons}$$

....

Underline the necessary information in each problem below. Write the proportions and solve the problems.

1. A slicing machine cut roast beef $\frac{1}{16}$ inch thick. The giant sandwich was advertised to contain roast beef 2 inches thick. How many slices of roast beef were on the sandwich?

2. The La Ronga Bakery baked 1,460 loaves of bread in 1 day. If each loaf contained $1\frac{3}{4}$ teaspoons of salt, how much salt did the bakery use?

3. How many books $\frac{7}{8}$ inch thick can be packed in a box 35 inches deep?

4. A can of pears weighs $9\frac{2}{3}$ ounces. There are 16 cans of pears in a carton. How many ounces does a carton of pears weigh?

5. There are 8 cups of detergent in a bottle of Easy Clean detergent. For one load of laundry, $\frac{1}{4}$ cup is all that is needed. How many loads of laundry can be cleaned with a bottle of Easy Clean?

Solving Conversion Word Problems

Have you ever seen this kind of problem?

EXAMPLE 1 Caren has a 204-**inch** roll of masking tape. How many **feet** of molding can she cover with the roll?

This problem is an example of a type of multiplication or division word problem that contains only one number and requires outside information in order to be solved. These are word problems involving **conversions** from one type of measurement to another.

Here is the solution and explanation of the example.

STEP 1 *question:* How many feet?

STEP 2 *necessary information:* 204 inches

Notice that the question asks for a solution that has a different label than what is given in the problem. To solve this, you must know how to convert inches to feet. Then you can set up a proportion to solve the problem. You may need to refer to the conversion chart on p. 219.

STEP 3 12 inches = 1 foot

labels for proportion: $\dfrac{\text{inches}}{\text{feet}}$

The conversion will be one side of the proportion.

$$\frac{12 \text{ inches}}{1 \text{ foot}} = \frac{204 \text{ inches}}{n \text{ feet}}$$

STEP 4 Cross multiply.

STEP 5 Divide.

$n = \textbf{17 feet}$

$$\frac{12}{1} \diagdown\!\!\!\!\diagup \frac{204}{n}$$

$$12 \times n = 204 \times 1$$
$$12n = 204$$
$$n = \frac{204}{12} = 17$$

A diagram can often help you picture a conversion word problem.

EXAMPLE 2 The Spring Lake Day-Care Center gives each of its 12 children a cup of milk for lunch every day. How many quarts of milk does the center use each day?

STEP 1 *question:* How many quarts of milk does the center use each day?

STEP 2 *necessary information:* 12 cups, 1 cup conversion formula: 4 cups = 1 quart

STEP 3 Decide what arithmetic operation to use.

The diagram shows that you should divide.

number of cups ÷ cups in a quart = number of quarts

STEP 4 Do the arithmetic.
$$12 \text{ cups} \div \frac{4 \text{ cups}}{1 \text{ quart}} = \textbf{3 quarts}$$

STEP 5 Make sure the answer is sensible. Look at the diagram to see that the answer of 3 quarts makes sense.

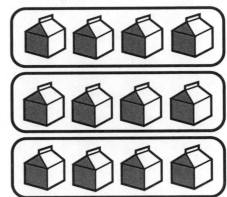

Use the conversion chart on page 219 to help you write the conversion and the proportion for each problem. Then solve the problem.

1. How many years old is Gloria's 30-month-old daughter?

2. A 200-gallon batch of ketchup was bottled in quart bottles. How many bottles were filled?

3. How many kilometers long is a 10,000-meter road race?

4. Paul's truck can carry a $\frac{1}{2}$-ton load. How many pounds of gravel can it carry?

5. José brought the cream shown at the right to the company picnic. How many ounces of cream did he bring?

6. Mt. Everest is 29,028 feet high. How many miles high is Mt. Everest? (Round to the nearest tenth.)

Using Proportions with Fractions and Decimals

Sometimes you will find a multiplication or division word problem in which one number is a decimal and the other is a fraction. The proportion method is very useful in solving this type of word problem.

EXAMPLE   A $\frac{3}{4}$-pound steak cost $6.75. How much did it cost per pound?

STEP 1   *question:* How much did it cost per pound?

STEP 2   *necessary information:* $\frac{3}{4}$ pound, $6.75

STEP 3   *labels for proportion:* pound, $

$$\frac{\frac{3}{4} \text{ pound}}{\$6.75} = \frac{1 \text{ pound}}{\$n}$$

STEP 4   Cross multiply.

STEP 5   Multiply both sides by $\frac{4}{3}$.

$$n = \$9.00$$

$$\frac{\frac{3}{4}}{6.75} \diagup\diagdown \frac{1}{n}$$

$$\frac{3}{4}n = 6.75$$

$$n = \overset{2.25}{\cancel{6.75}} \times \frac{4}{\cancel{3}_1}$$

$$n = 9.00$$

If you had not been able to cancel in the example, you would have multiplied the numerators and divided by the product of the denominators. Keep the decimal point in the correct place.

Write the proportions and solve the problems below. Round money problems to the nearest cent.

1. Cloth was being sold at $12.60 a yard. Lori bought $3\frac{1}{3}$ yards of cloth. How much did she spend?

2. George was told that 13.5 pounds of time-release fertilizer should last $4\frac{1}{2}$ years. How much fertilizer is used each year?

3. Murray bought $2\frac{2}{3}$ pounds of grapes for $3.25. How much did the grapes cost per pound?

4. Marty took pictures at graduation. He used $7\frac{1}{2}$ rolls of film and was charged $384.50. What was the charge per roll of film?

Using Proportions with Multiplication and Division

Underline the necessary information. Write the proportions and solve the problems.

1. A butcher can cut up a chicken in $\frac{1}{12}$ of an hour. How many chickens can be cut up in an 8-hour work day?

2. A nurse can take 8 blood samples in 60 minutes. How long does it take her to take one blood sample?

 3. A mile is about 1.6 kilometers. How many kilometers is a 26-mile marathon?

4. An oil-drilling rig can drill 6 feet in an hour. How far can it drill in 24 hours?

 5. A gram is 0.04 ounces. How many grams are in a 12-ounce can of pineapple juice?

6. Rose uses $3\frac{1}{4}$ pounds of pumpkin to make 2 pumpkin pies. For a fall bake sale, she made 10 pies. How many pounds of pumpkin did she use?

 7. When the floodgates were opened, 68,000 gallons of water flowed over the dam per hour. How many gallons flowed over the dam in a day?

8. Lace trimming costs $0.12 per foot. How much did Zelda spend on $4\frac{1}{4}$ feet of trimming?

9. Super Glue sets in $3\frac{1}{2}$ minutes. In how many seconds does Super Glue set?

10. A 942-page book contained 302,382 words. On the average, how many words were on each page?

STRATEGIES WITH MIXED WORD PROBLEMS

Mixed Word Problems with Whole Numbers

So far, you have worked with word problems that have been divided into two major categories—addition/subtraction problems and multiplication/division problems.

In most situations, you will be faced with the four types of problems mixed together. Always read each problem carefully to get an understanding of the situation it describes. This will help you choose the right arithmetic operation.

Keep these general guidelines in mind:

- when combining amounts   → add

- when finding the difference between two amounts   → subtract

- when given one unit of something and asked to find several   → multiply

- when asked to find a fraction of a quantity   → multiply

- when given the amount for several and asked for one   → divide

- when dividing, cutting, sharing   → divide

Working through the following exercises will help sharpen your skills with word problems when the different types are mixed together.

Write the arithmetic operation (addition, subtraction, multiplication, or division) that you would use to solve each problem. DO NOT SOLVE!

1. Doreen needs 39 credits to complete her bachelor's degree at the state university. The university charges $265 per credit for tuition. How much will Doreen have to pay in tuition if she completes her degree?

2. Alan wrote a 74,200-word manuscript for his new book. The typesetter estimates that there will be an average of 280 words per page. If the typesetter is correct, how many pages will there be in the book?

3. After laying off 27 workers, Paul still had 168 workers at the hospital. How many workers were there at the hospital before the changes?

4. A department store bought shirts for $14 each and sold them for $24. How much profit did it make on each shirt?

5. Tom needs 19 feet of molding for each doorway in his home. The home will have 9 doorways. How much molding does he need?

6. Part of Rob's harvesting log is shown at the right. How many more ears of corn did he harvest on Thursday than on Wednesday?

	WED	THURS
Corn	476	548
Zucchini	94	129

7. As coach of her soccer team, Althea decided that all 22 players would get equal playing time. With a total of 990 minutes to distribute, how many playing minutes did Althea give to each of her players?

8. At her day-care center, Beth used an entire gallon of juice at snack time for 24 children. On the average, how many ounces of juice did each child receive?

9. The governor's goal is to reduce the state's imports of foreign oil by 25,000 barrels to 90,000 barrels a month. How much oil is the state currently importing a month?

Using Labels to Solve Word Problems

Every number in a word problem has a **label.** Every number *refers* to something. In other words, it makes no sense to say simply "7" or "$38\frac{1}{2}$." We need to know—*7 what? $38\frac{1}{2}$ what?* Dogs? Miles per hour? Years old? What do these numbers refer to?

Paying careful attention to labels will help you decide whether to add, subtract, multiply, or divide. Look at the following example.

EXAMPLE 1 On Saturday night, Bruce spent $46.50 on dinner and $38.00 for tickets to a play. How much did he spend altogether?

Notice that the labels of both pieces of the necessary information are *dollars.* Also, you can tell that the label of the answer will be in *dollars.* You probably already have figured out that you need to add $46.50 and $38.00 to solve the problem.

Look at the next example.

EXAMPLE 2 In the last election, 35,102 women and 29,952 men voted. How many people voted?

The labels of the necessary information and the answer are different. Does this mean you should not add or subtract?

Whenever the labels of items in a word problem are different, first ask yourself if the different labels can be part of a *broader category* or if they can be *converted* to a common unit (such as from *pounds* to *ounces* or from *years* to *months*). For example, *men* and *women* can both be considered part of the broader category of *people*; therefore, all the labels in the problem are the same and you can add or subtract to find the answer. For the problem above, you should add 35,102 and 29,952 to get the answer.

Using labels to decide whether to multiply or to divide is a bit trickier. However, if you are willing to play a bit with the labels in a problem, you can often make this decision before you have to work with actual numbers. Read the next example.

> You will often find this pattern in word problems:
>
> When the labels of all the necessary information and the answer are the same, you usually need to *add* or *subtract* to solve.

EXAMPLE 3   Ken drove 385 miles in 7 hours. How many miles per hour did he average on this trip?

The labels of one piece of the necessary information is *miles*; the label of the other piece is *hours*. The label of the answer is *miles per hour*. Already you may be guessing that you should not add or subtract, for the labels are *not* the same and they can't be converted to a common unit.

The answer will be in *miles per hour*, which can also be written as $\frac{\text{miles}}{\text{hour}}$ (miles *divided by* 1 hour). Set up a statement using the labels from the problem:

$$\text{miles} \; \Box \; \text{hours} = \frac{\text{miles}}{\text{hour}}$$

> **Remember:** *miles per hour* means the ratio
> $$\frac{\text{miles}}{\text{hour}}$$

How would you fill in the box? Ask yourself, "What do I have to do to *miles* and *hours* in order to get $\frac{\text{miles}}{\text{hour}}$?"

You need to divide.

$$385 \text{ miles} \; \boxed{\div} \; 7 \text{ hours} = \frac{\textbf{55 miles}}{\textbf{hour}}$$

OR

$$385 \text{ miles} \div 7 \text{ hours} = \textbf{55 miles per hour}$$

Now look at the next example.

EXAMPLE 4   Ken took a 7-hour trip. He averaged 55 miles per hour on the trip. How many miles in all did he drive?

The label of one piece of the necessary information is *hours*; the label of the other piece is *miles per hour*. The label of the answer will be *miles*. Ask yourself, "What would I do to *hours* and *miles per hour* to get *miles*?"

$$\text{hours} \; \Box \; \frac{\text{miles}}{\text{hour}} = \text{miles}$$

Try multiplying. Just as with numerical multiplication, you can cancel out common factors (in this case, the label *hour*).

canceling *using words*

$$\cancel{\text{hours}} \; \boxed{\times} \; \frac{\text{miles}}{\cancel{\text{hour}}} = \text{miles}$$

canceling *using numbers*

$$\cancel{5} \times \frac{3}{\cancel{5}} = 3$$

Because canceling labels leaves you the label to your answer, you know that you should multiply to get the correct answer.

$$\cancel{\text{hours}}\ \square\ \frac{\text{miles}}{\cancel{\text{hour}}} = \text{miles}$$

$$7\ \cancel{\text{hours}}\ \boxed{\times}\ \frac{55\ \text{miles}}{1\ \cancel{\text{hour}}} = ?\ \text{miles}$$

$$7 \times 55\ \text{miles} = \textbf{385 miles}$$

Now, look at a problem in which you can't cancel the labels.

EXAMPLE 5 Ken took a 385-mile trip and drove 55 miles per hour. How many hours did he drive?

Look at the expression. $\text{miles}\ \square\ \dfrac{\text{miles}}{\text{hour}} = \text{hours}$

Can you convert the labels to a common unit? No. Can you cancel the labels *miles* and *hours*? No. You have one more case you should try.

Remember the procedure for dividing fractions is to invert (flip over) the fraction you are dividing by and then multiply the result. You can do the same thing with labels.

$$385\ \text{miles}\ \boxed{\div}\ \frac{55\ \text{miles}}{1\ \text{hour}} = ?\ \text{hours}$$

$$385\ \cancel{\text{miles}} \times \frac{1\ \text{hour}}{55\ \cancel{\text{miles}}} = ?\ \text{hours}$$

$$385 \times \tfrac{1}{55}\ \text{hours} = 7\ \text{hours}$$

When you actually do the math, it is easier just to divide, but it is important to know that you can still use the labels to decide which operation to use.

....

Read each of the following problems. Look at the labels and decide if they can be renamed to a broader category. Then write in the correct operation to solve the problem. Finally, solve the problem.

1. Sam cut an 8-ounce slice from a 20-pound round of cheese. How many ounces of cheese were left?

    necessary information labels: _____ _____

    answer label: _____

    Are the labels different? _____*yes*_____

    If so, what is the new label? __*ounces*__

    _____ $\square$ _____ = _____
       label   label   label

    Answer: _____

2. Adrienne needed to cut 2 feet from a 72-inch piece of molding. After the cut, how much molding was left?

necessary information labels: _____ _____

answer label: _____

Are the labels different? _____

If so, what is the new label? _____

_____ ▢ _____ = _____
    label   label   label

Answer: _____

3. Maura's hair was $9\frac{1}{2}$ inches long. How long was her hair before she had cut it by $1\frac{3}{4}$ inches?

necessary information labels: _____ _____

answer label: _____

Are the labels different? _____

If so, what is the new label? _____

_____ ▢ _____ = _____
    label   label   label

Answer: _____

Read the following problems. First look at the labels in the problem and see if they can be canceled. If the labels can't be canceled, flip over the second label, and then check if the labels can be canceled. Then write the correct operation in the box. Finally, solve the problem.

4. A ream of paper contains 500 sheets. A box contains 10 reams of paper. How many sheets of paper are in the box?

$\dfrac{\text{sheets}}{\text{ream}}$ ▢ reams = sheets   Answer: _____

5. It cost $6 to go to the movies. A movie theater collected $522 in ticket sales. How many tickets were sold?

dollars ▢ $\dfrac{\text{dollars}}{\text{ticket}}$ = tickets   Answer: _____

6. An average tomato plant in John's garden yields 8 pounds of tomatoes. How much can he expect from his 14 plants?

plants ▢ $\dfrac{\text{pounds}}{\text{plant}}$ = pounds   Answer: _____

Not Enough Information

Now that you know how to decide whether to add, subtract, multiply, or divide to solve a word problem, you should be able to recognize a word problem that cannot be solved because not enough information is given.

Look at the following example.

EXAMPLE 1   At her waitress job, Sheila earns $4.50 an hour plus tips. Last week she earned $65.40 in tips. How much did she earn last week?

STEP 1   *question:* How much did she earn last week?

STEP 2   *necessary information:* $4.50/hour, $65.40

STEP 3   Decide what arithmetic operation to use.

tips + (pay per hour × hours worked) = total earned

missing information: hours worked

At first glance, you might think that you have enough information since there are two numbers. But when the solution is set up, you can see that you need to know the number of hours Sheila worked to find out what she earned.

For each word problem, circle the letter of the information needed to solve the problem.

1. A supermarket sold 350 pounds of bananas at $0.59 a pound. How many pounds of bananas did it have left?

   a. You need to know how much the supermarket paid for the bananas.
   b. You need to know how many pounds of bananas the supermarket started with.
   c. You need to know how much money the supermarket made for each pound of bananas sold.
   d. You have enough information to solve the problem.

2. In one day last year, 2,417 people were born or moved into the state and 1,620 people died or left the state. What was the state's change of population for the day?

   a. You need to know the total population of the state.
   b. You need to know the name of the state.
   c. You need to know exactly how many people were born and exactly how many died.
   d. You have enough information to solve the problem.

3. Roast beef that normally costs $2.59 a pound was marked down $0.60. If Gina paid for a roast with a $10.00 bill, how much change did she receive?

   a. You need to know the weight of Gina's roast beef.
   b. You need to know the total amount of money Gina had.
   c. You need to know how much the supermarket paid for the roast beef.
   d. You have enough information to solve the problem.

4. A loaded truck carrying boxes of books weighed 7,105 pounds at the weigh station. If each box of books weighed 42 pounds, how much did the unloaded truck weigh?

   a. You need to know how many books were in the truck.
   b. You need to know the weight of a single book.
   c. You need to know how many boxes were in the truck.
   d. You have enough information to solve the problem.

5. A bag of 40 snack bars weighs 12 ounces. How much does each snack bar weigh?

   a. You need to know the price of one snack bar.
   b. You need to know how many ounces are in a pound.
   c. You need to know the total price of the entire bag.
   d. You have enough information to solve the problem.

In the following problems, decide whether to add, subtract, multiply, or divide. Then solve the problem. Circle the letter of the correct answer.

6. Mr. Gomez's obituary appeared in a 1998 newspaper. It said that he was 86 years old when he died and had been married for 51 years. In what year was he born?

   a. 1903
   b. 1947
   c. 1861
   d. 1887
   e. 1912

7. An electrician has a piece of cable the length shown at the right. How long a cable would he have if he laid 7 of these pieces end to end?

   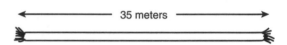

   a. 5 meters
   b. 28 meters
   c. 42 meters
   d. 245 meters
   e. 490 meters

8. The population of San Jose rose by 37,250 people. The population had been 782,250. What was the new population?

   a. 21 times
   b. 745,000 people
   c. 819,500 people
   d. 29,138,812,500 people
   e. not enough information given

9. A Christmas light uses 2 watts of electricity. How many lights can be strung on a circuit that can handle a load of 300 watts?

   a. 150 lights
   b. 600 lights
   c. 298 lights
   d. 302 lights
   e. none of the above

10. A telephone cable can handle 12,500 calls at any one time. How many cables are needed to handle a peak load of 87,500 calls?

  a. 100,000 cables
  b. 75,000 cables
  c. 7 cables
  d. 70 cables
  e. none of the above

11. Gene bought $360 worth of sports equipment and $18 worth of office supplies for the boys' club. Since the boys' club is tax-exempt, he didn't have to pay the sales tax. If he had paid tax, how much would he have spent?

  a. $20
  b. $378
  c. $342
  d. $360
  e. not enough information given

12. A clothing factory produced 8,760 yards of cloth. What was the average production from each of the 60 looms in the factory?

  a. 146 yards
  b. 1,460 yards
  c. 8,700 yards
  d. 8,820 yards
  e. 525,600 yards

13. Len's goal was to sell 20 encyclopedias a month. Part of Len's sales log is shown at the right. By how much did he exceed his goal in September?

  a. 57 encyclopedias
  b. 17 encyclopedias
  c. 13 encyclopedias
  d. 2 encyclopedias
  e. 740 encyclopedias

Encyclopedia Sales	
August	19
September	37
October	30
November	21

In the following problems, write a question that matches the solution.

14. In 1996 the per capita income, the average amount of money each person earned, in the District of Columbia was $29,202. Per capita expenses, the average amount of money each person spent, was $26,097.

    $\$29,202 - \$26,097 = n$

15. During the pre-Christmas sale, the price of the new car was slashed from $15,364 to $11,994.

    $\$15,364 - \$11,994 = n$

16. In order to be stained, concrete needs to be treated with 4 parts muriactic acid mixed with 1 part water. Jaresh had a 1 pint container of muriactic acid.

    $$\frac{4 \text{ parts muriactic acid}}{1 \text{ part water}} = \frac{1 \text{ pint muriactic acid}}{n \text{ pints water}}$$

17. When building a roof, Rosalia knew that she needed 1 square foot of ventilation for 300 square feet of attic. Her attic was 1,200 square feet.

    $$\frac{1 \text{ sq ft ventilation}}{300 \text{ sq ft attic}} = \frac{n \text{ sq ft ventilation}}{1,200 \text{ sq ft attic}}$$

18. A base coat of stucco uses $2\frac{1}{2}$ parts clean common sand to 1 part Type N cement. Cesar has a 50-pound bucket of Type N cement.

    $$\frac{2\frac{1}{2} \text{ parts sand}}{1 \text{ part Type N cement}} = \frac{n \text{ lb sand}}{50 \text{ lb Type N cement}}$$

19. A drill press can drill a hole accurately to 0.002 inch. The press was set to drill a 0.235 hole.

    $0.235 \text{ inch} + 0.002 \text{ inch} = n \text{ inch}$

20. Abad and Socorro stayed overnight at Motel 5. The room was $39.95. The motel taxes were $6.25.

    $\$39.95 + \$6.25 = n$

Mixed Word Problems—Whole Numbers, Decimals, and Fractions

**In the following problems, circle the letter of the correct answer.
Round decimals to the nearest cent or the nearest hundredth.**

1. Sandy bought a roast beef sandwich for $1.89, which included
   $0.09 tax. What was the cost of the sandwich alone?

   a. $1.89
   b. $1.80
   c. $1.98
   d. $2.10
   e. $1.70

2. In 1994 the estimated population of the United States was
   260,714,000. The 1994 estimated population of Russia was
   149,609,000. How much greater was the population of the
   United States than the population of Russia?

   a. 111,105,000 people
   b. 410,323,000 people
   c. 129,115,000 people
   d. 309,313,000 people
   e. none of the above

3. To tie her tomato plants, Emmy cut the string shown at the
   right into $\frac{3}{4}$-foot-long pieces. How many pieces of string
   did she have to tie her tomatoes?

   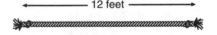

   a. $12\frac{3}{4}$ feet
   b. $11\frac{1}{4}$ feet
   c. 9 pieces
   d. 16 pieces
   e. none of the above

4. Boneless chicken breasts cost $1.95 a pound. Cali paid $8.70 for
   a package of chicken breasts. How much did the chicken breasts
   weigh? (Round to nearest hundredth.)

   a. 6.75 pounds
   b. 16.97 pounds
   c. 10.65 pounds
   d. 4.46 pounds
   e. 2.24 pounds

5. At the New York Stock Market, a stock opened at $20\frac{3}{8}$ a share and closed at the end of the day at $22\frac{1}{2}$. How much did it gain for the day?

    a. $42\frac{7}{8}$

    b. $2\frac{1}{8}$

    c. $2\frac{1}{3}$

    d. $2\frac{2}{5}$

    e. none of the above

6. What is the total weight of the two packages of fruit shown at the right?

    a. 1.81 pounds
    b. 1.63 pounds
    c. 2.62 pounds
    d. 1.55 pounds
    e. 0.82 pound

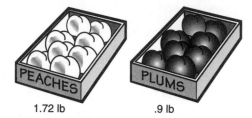

PEACHES 1.72 lb   PLUMS .9 lb

7. A serving of Kellogg's Raisin Bran contains 0.26 gram of potassium. How many grams of potassium are in an 11-serving package of Raisin Bran?

    a. 11.26 grams
    b. 10.74 grams
    c. 42.31 grams
    d. 2.86 grams
    e. 0.02 gram

8. The auto repair shop charged Muriel $1,125 to repair her car. She had a $250-deductible insurance policy. How much did the insurance company pay for the repair of her car?

    a. $1,375.00
    b. $875.00
    c. $4.50
    d. $281.25
    e. none of the above

9. Debbie spent $\frac{1}{3}$ of her paycheck on food and $\frac{1}{4}$ for clothes. Her paycheck was for $414. How much did she spend for food?

   a. $138.00
   b. $34.50
   c. $241.50
   d. $1,242.00
   e. $103.50

10. State Airlines does a complete maintenance check of its airplanes every 12,000 miles flown. Airplane #200 was flown 96,000 miles last year. How many complete maintenance checks did it have last year?

    a. 480 maintenance checks
    b. 84,000 miles
    c. 108,000 miles
    d. 8 maintenance checks
    e. 1,152,000,000 miles

11. After getting a tune-up, Ernie was able to drive 283.1 miles on 14.9 gallons of gas. How many miles did he get per gallon?

    a. 19 miles
    b. 42.18 miles
    c. 134 miles
    d. 298 miles
    e. 268.2 miles

12. The Platte River is normally 7 feet deep. During a recent flood, it crested at 14 feet above normal. What was the depth of the river at the crest of the flood?

    a. 7 feet
    b. 21 feet
    c. 98 feet
    d. 2 feet
    e. none of the above

13. The *Concorde* flew 3,855 miles across the Atlantic in $3\frac{3}{4}$ hours. What was its average speed?

    a. $3,858\frac{3}{4}$ miles per hour
    b. $3,851\frac{1}{4}$ miles per hour
    c. 1,028 miles per hour
    d. $14,456\frac{1}{3}$ miles per hour
    e. $467\frac{9}{33}$ miles per hour

PERCENT WORD PROBLEMS

Identifying the Parts of a Percent Word Problem

Read the statements below.

The 8-ounce glass is 50% full. It contains 4 ounces.

These statements contain three facts:

the whole: the 8-ounce glass

the part: 4 ounces

the percent: 50%

A one-step percent word problem would be missing one of these facts. When you are solving a percent word problem, first identify what you are looking for. As shown above, you have three possible choices: *the part, the whole,* or *the percent.*

It is usually easiest to figure out that you are being asked to find the percent. Word problems asking for the percent usually ask for it directly, with a question such as "What is the percent?" or "Find the percent" or "Three is what percent?" Occasionally, other percent-type words are used, such as "What is the *interest rate*?"

EXAMPLE 1   What percent of 30 is 6?

The question asks *is what percent?* Therefore, you are looking for the percent.

Sometimes you are given the percent and one other number. You must decide whether you are looking for the part or the whole.

EXAMPLE 2   81% of what number is 162?

The phrase *of what number* means you are looking for the whole.

EXAMPLE 3 Yesterday 114 city employees were absent. This was 4% of the city work force. How many people work for the city?

  STEP 1 *question:* How many people work for the city?

  STEP 2 *necessary information:* 114 city employees, 4%

  STEP 3 You are given the number of city employees who were absent (114) and the percent of the work force that this represents (4%). You are looking for the total number of people who work for the city, the whole.

EXAMPLE 4 What number is 75% of 40?

  You are looking for a number that is a percent of another number. You are looking for the part.

EXAMPLE 5 Operating at full capacity, the automobile plant produced 25 cars an hour. How many cars did the plant produce when operating at 40% capacity?

  STEP 1 *question:* How many cars did the plant produce?

  STEP 2 *necessary information:* 25 cars, 40%

  STEP 3 You are given the production at full capacity (25 cars an hour). To find the production at 40% capacity, you solve for the part.

....

For each problem, write down whether you are looking for the part, the whole, or the percent. DO NOT SOLVE!

  1. The city reported that 14,078 out of 35,817 registered voters voted in the election. What percent of registered voters voted in the election?

  2. A total of 14,615 people voted in the election. The election results are shown at the right. How many votes did the winning candidate get?

Vote Percentages	
Candidate A	54%
Candidate B	39%
Candidate C	7%

3. An ad said that 36% of the plumbers polled recommended Drāno. If 72 plumbers recommended Drāno, how many plumbers were polled?

4. Eric found that 85% of a roll of 36 pictures were perfect prints. How many perfect prints did he get from the roll?

5. A seed company guaranteed 87% germination of its spinach seed. If Jed had 450 spinach seeds germinate, how many seeds did he plant?

6. The state had a work force of 1,622,145. If 132,998 of these people were unemployed, what was the unemployment rate for the state?

7. A bedroom set normally priced at $1,400 is on sale. How much would Rochelle save if she bought the set at the advertised sale, shown at the right, instead of at the regular price?

8. Last year, 980 people took the high school equivalency exam at the local official test center. If 637 people passed the exam, what percent of the people taking the exam passed?

9. If 8% of the registered voters sign the initiative petition, it will be placed on the November ballot. There are 193,825 registered voters in the county. How many of them must sign the petition for it to go on the ballot?

10. An independent study group estimated that only 35% of all crimes in the city were reported. Last year 2,800 crimes were reported. According to the study, how many crimes were actually committed?

Solving Percent Word Problems

Once you identify what you are looking for in a percent word problem, set up the problem and solve it.

Percent word problems can be solved using proportions. These problems can be set up in the following form:

$$\frac{\text{part}}{\text{whole}} = \frac{\%}{100}$$

This proportion means that the ratio of the part to the whole is equal to the ratio of the percent to 100.

Using the proportion method, you can solve for the part, the whole, or the percent. The percent is always written over 100 because the percent represents a fraction with 100 in the denominator.

As you saw in your earlier work with proportions, a proportion is the same as two equivalent fractions. For example,

2 is 50% of 4 and can be written as

$$\frac{2}{4} = \frac{50\%}{100}$$

2 is the *part*, 4 is the *whole*, and 50 is the *percent*.

EXAMPLE 1   4 is what percent of 16?

STEP 1   *question:* is what percent?

You are looking for the percent.

STEP 2   *necessary information:* 4 is, of 16

For this type of percent exercise, the word *is* follows the part, and the number after *of* is the whole.

STEP 3   Set up a proportion in this form:
numbers   *percents*

$$\frac{\text{part}}{\text{whole}} = \frac{\text{percent}}{100}$$

Fill in the proportion with the given information from the problem. Call the number you are looking for *n*.
numbers   *percents*

$$\frac{4}{16} = \frac{n}{100}$$

STEP 4   Cross multiply.

$$16 \times n = 4 \times 100$$
$$16n = 400$$

STEP 5   Divide.

$$n = \frac{400}{16} = \boldsymbol{25\%}$$

EXAMPLE 2   If 24 out of 96 city playgrounds need major repairs, what percent of the city playgrounds need major repairs?

     STEP 1   *question:* What percent of the city playgrounds need major repairs?

                 You are looking for the percent.

     STEP 2   *necessary information:* 24 out of 96

                 96 is the whole (all the playgrounds).

                 24 is the part (playgrounds needing repairs).

     STEP 3   <u>numbers</u>   <u>percents</u>

$$\frac{24 \text{ playgrounds}}{96 \text{ playgrounds}} = \frac{n}{100}$$

     STEP 4   Cross multiply.

$$96 \times n = 24 \times 100$$
$$96n = 2{,}400$$

     STEP 5   Divide.

$$n = \frac{2{,}400}{96} = \boldsymbol{25\%}$$

EXAMPLE 3   30% of what number is 78?

     STEP 1   *question:* of what number?

                 You are looking for the whole.

     STEP 2   *necessary information:* 30%, is 78

                 30% is the percent.

                 78 is the part.

     STEP 3   Set up a proportion.

$$\frac{78}{n} = \frac{30}{100}$$

     STEP 4   Cross multiply.

$$30 \times n = 78 \times 100$$
$$30n = 7{,}800$$

     STEP 5   Divide.

$$n = \frac{7{,}800}{30} = \boldsymbol{260}$$

EXAMPLE 4 The finance company required that Lynn make a down payment of 15% on a used car. She can afford a down payment of $600. What is the most expensive car that she could buy?

STEP 1 *question:* What is the most expensive car that she could buy?

You are looking for the whole (the price of the car).

STEP 2 *necessary information:* 15%, $600

15% is the percent.
$600 is the down payment, which is a part of the total price of the car.

STEP 3 Set up a proportion.

$$\frac{\$600}{\$n} = \frac{15}{100}$$

STEP 4 Cross multiply.

$$15 \times n = 600 \times 100$$
$$15n = 60,000$$

STEP 5 Divide.

$$n = \frac{60,000}{15} = \$4,000$$

EXAMPLE 5 What is 40% of 65?

STEP 1 *question:* What is?

You are looking for the part.

STEP 2 *necessary information:* 40%, of 65

40% is the percent.

65 is the whole.

STEP 3 Set up a proportion.

$$\frac{n}{65} = \frac{40}{100}$$

STEP 4 Cross multiply.

$$100 \times n = 65 \times 40$$
$$100n = 2,600$$

STEP 5 Divide.

$$n = \frac{2,600}{100} = 26$$

EXAMPLE 6 June decided that she could spend 25% of her income for rent. She makes $1,740 a month. How much can she spend for rent?

STEP 1 *question:* How much can she spend for rent?

You are looking for the part of her income that she will spend on rent.

STEP 2 *necessary information:* 25%, $1,740

25% is the percent.
$1,740 is her whole income.

STEP 3 Set up a proportion.

$$\frac{\$n}{\$1,740} = \frac{25}{100}$$

STEP 4 Cross multiply.

$$100 \times n = 1,740 \times 25$$
$$100n = 43,500$$

STEP 5 Divide.

$$n = \frac{43,500}{100} = \$435$$

Use proportions to solve the following problems.

1. 36 is what percent of 144?

2. 288 is 72% of what number?

3. What is 68% of 75?

4. A $160 suit was reduced by $40. What was the percent of the reduction?

5. The election results are shown at right. If 28,450 votes were cast in the school board election, how many votes did Marsha receive?

6. The state government cut aid for adult education by 25%. Metropolis expects to lose $96,000. How much aid for adult education had Metropolis been receiving?

7. Last year Jeffrey paid 7% of his income in taxes. He paid $1,659. What was his income?

School Board Elections	
Clayton	19%
Andrea	21%
Marcus	2%
Marsha	58%

8. Exactly 60% of the residents of the city are African American. The population of the city is 345,780. How many African American people live in the city?

9. In 1998, Robyn paid $340 interest on $2,000 that she had borrowed. What was the interest rate on the borrowed money?

10. In order to control his chloresterol, Sespend tried to limit his calories from fat to 20% of his total calories. On an average day, he eats 2,400 calories. What is the most calories from fat he can eat on an average day if he is to meet his goal?

For each word problem, select the correct question that matches the solution.

11. Khan had a special coupon that allowed her to take 20% off the price of a clearance item. She decided to buy a sweater she wanted that had a clearance price of $16.

   a. How much did she pay for the sweater?
   b. How much money did she have left?
   c. How much did she save with her coupon?
   d. What was the original price of the sweater?
   e. How many sweaters could she buy?

$$\frac{\$n}{\$16} = \frac{20\%}{100}$$

$$n = \frac{320}{100} = \$3.20$$

12. A national survey found that 720 out of 900 dentists recommended using Never Breaks dental floss.

   a. How many dentists recommended a different brand?
   b. How many more dentists recommended Never Break than all other brands?
   c. How many different brands were recommended?
   d. What percent of dentists recommended other brands?
   e. What percent of dentists recommended Never Break?

$$\frac{720 \text{ dentists}}{900 \text{ dentists}} = \frac{n\%}{100}$$

$$900n = 72,000$$

$$n = \frac{72,000}{900} = 80\%$$

13. The state passed a law requiring 60% of legislators to vote for a tax increase in order for it to pass. The lower house had 180 members.

  a. At least how many legislators in the lower house had to support a tax increase in order for it to pass?
  b. At least how many legislators in the lower house had to oppose a tax increase in order for it not to pass?
  c. What was the largest number of legislators that could oppose a tax increase that passes in the lower house?
  d. How many more legislators are needed to pass a tax increase in the lower house than when only a 50% majority was needed?
  e. What was the smallest possible margin of victory?

$$\frac{n \text{ members}}{180 \text{ members}} = \frac{60\%}{100}$$

$$100n = 10{,}800$$

$$n = 108 \text{ legislators}$$

14. Floor Mart advertised a total savings of 30% off list price for their store brand microwave oven. When Wilson bought the microwave oven, he saved $75 from the list price.

  a. What percent of the list price did he pay?
  b. What was the list price of the oven?
  c. What was the sale price of the oven?
  d. How much change did Wilson receive?
  e. How much more would he have paid if he had paid full price?

$$\frac{\$75}{\$n} = \frac{30\%}{100}$$

$$30n = 7500$$

$$n = \frac{7500}{30} = \$250$$

15. The production line for the molded plastic parts needs to be stopped and the machines readjusted if 2% or more of the parts are rejected by the inspectors. In an hour, 600 parts are made on the production line.

  a. What is the largest number of parts that can be defective in an hour?
  b. What is the smallest number of parts that can be defective in an hour?
  c. What is the largest possible number of good parts that can be made in an hour in which the production line needs to be stopped?
  d. What is the smallest possible number of defective parts that need to be found in an hour in order to shut down the production line?
  e. How many more good parts than rejected parts were produced?

$$\frac{n \text{ parts}}{600 \text{ parts}} = \frac{2\%}{100}$$

$$100n = 1200$$

$$n = 12 \text{ parts}$$

Percent and Estimation

Have you ever listened to a report of election results? The reporter will often say something like, "The Congressman was reelected with 54% of the vote." This percent is an estimate, not an exact amount. The reporter has rounded the result to the nearest percent.

 In each situation described, decide whether it is more likely that the percent given is an estimate or an exact amount.

1. The state unemployment rate was 5.3% last month.

   a. exact
   b. estimate

2. During the sale, all clothing prices are reduced 20%.

   a. exact
   b. estimate

3. Scientists say that there is a 15% chance of a major earthquake in the region in the next ten years.

   a. exact
   b. estimate

4. Even 20 years after the last underground nuclear tests, background radiation levels were 35% above normal.

   a. exact
   b. estimate

5. Angel's son Alberto got a grade of 88% correct on the 50-question multiple-choice test.

   a. exact
   b. estimate

6. In the city, 24% of the people were living in poverty.

   a. exact
   b. estimate

For each word problem, find a solution that could be true. Some questions have a range of possible correct answers. Others have only one possible correct answer.

7. The newscast reported that Elena received 65% of the vote in the local election in which 4,864 votes were cast. How many votes might she have received?

8. The official report said that the city population was expected to increase 5% in the next decade. The current population is 128,431. What could the population be in a decade and still be within the predicted range?

9. The investment analyst predicted a 12% return on the mutual fund. Sugi invested $3,000 and at the end of the year had earned $352 on that investment. Was the prediction accurate?

10. The state printed 20,000,000 Instant Winner Lottery Tickets and claimed that 3% of the tickets were winners, with the prizes ranging from a free ticket to $100,000. How many winning tickets were printed?

11. An Internet provider predicted that its number of customers would increase 18% in a year. At the start of the year, the company had 56,820 customers. At the end of the year, it had 69,965 customers. Was the Internet provider's prediction accurate?

The Percent Circle

If you find working with proportions to solve percent problems too abstract, you could use a memory aid called the **percent circle.** The percent circle is equivalent to a proportion, but it creates a picture to help you decide whether to multiply or divide.

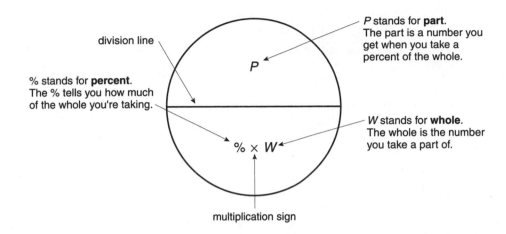

Using the Percent Circle

EXAMPLE 1 Finding a part of the whole.

According to Francisco's union contract, he is due to get a 4% raise. He currently earns $350 a week. What will his raise be?

STEP 1 The raise is calculated as a part of his salary. Therefore cover P (the part) since that is the number you're trying to find.

STEP 2 Read the uncovered symbols: % × W.

To find the part, multiply the percent by the whole.

STEP 3 Write the percent as either a fraction or a decimal.
$$4\% = \frac{4}{100} \text{ or } 0.04$$

STEP 4 Do the calculation.

$$\frac{4}{100} \times 350 = \frac{1400}{100} = \$14 \text{ or } 0.04 \times 350 = \$14.00$$

Francisco will get a **$14 raise.**

Note: When you use the percent circle, you must remember that *percent* means "compared to 100." When you know the percent, remember to divide it by 100. When you are looking for the percent, remember to multiply by 100.

EXAMPLE 2 Finding a whole when a part and a percent are given.

The Cheap Cars used car lot requires buyers to pay 20% of the total price of a car as a down payment. The buyer can finance the rest of the purchase price. Celso can spend up to $800 on a down payment for a used car. What is the highest price used car he could buy at Cheap Cars?

STEP 1 Since the down payment is a part of the price of a car, you are trying to find the whole. Therefore, cover W (the whole) since that is the number you're trying to find.

STEP 2 Read the uncovered symbols. $\frac{P}{\%}$ means $P \div \%$.

To find the whole, divide the part by the percent.

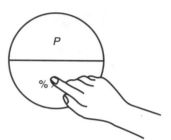

STEP 3 Write the percent as either a fraction or a decimal.

$$20\% = \frac{20}{100} = \frac{1}{5} \text{ or } 0.20$$

STEP 4 Do the calculation.

$$800 \div \frac{1}{5} = 800 \times \frac{5}{1} = \$4,000, \text{ or } 20\overline{)800.00}^{\;\$4000}$$

The highest price used car Celso could buy at Cheap Cars would cost **$4,000.**

EXAMPLE 3 Finding what percent a part is of a whole.

A clinic tested 360 drug abusers for HIV and found that 284 tested positive. What percent of the drug abusers tested HIV positive?

STEP 1 You need to find what percent 234 is of 360.
You are looking for the percent. Therefore cover %.

STEP 2 Read the uncovered symbols. $\frac{P}{W}$ means $P \div W$.

To find the percent, divide the part by the whole.

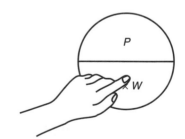

STEP 3 Do the calculation.

$$\frac{234}{360} \text{ or } 360\overline{)234.00}^{\;0.65}$$

STEP 4 Convert the decimal to a percent.

$$0.65 \times 100 = 65\%$$

65% of the drug abusers tested HIV positive.

Note: It is possible for the part to be larger than the whole. If the part is larger than the whole, the percent will be larger than 100%.

Use the percent circle to solve the following word problems.

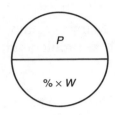

1. Dinner cost Luis $40. He wanted to leave the waitress a 15% tip. How much should he leave for a tip?

2. Armand bought a chain saw for $80. How much did he pay in sales tax if the sales tax was 5%?

3. Webmasters Computer Company has 540 employees. Of those employees, 27 are over age 60. What percent of the employees are over age 60?

4. Super Foods' market research showed that 2% of all households that received a weekly circular actually came to shop at the store. The store manager of a new Super Foods planned for 1,500 different customers to visit the Super Foods store in its first week. How many circulars should she plan to send out?

5. Ninety days after the end of training, 24 out of 30 graduates of the computer repair training program had found jobs. What was the job placement rate of the computer repair training program?

6. The Serious Disease Charity spent $28,000 to raise $112,000 in contributions. What percent of the money raised was spent on fund raising?

Solving Percent Word Problems with Decimals and Fractions

Many percent word problems also contain decimals or fractions. These problems are also solved using the proportion method.

EXAMPLE 1   What is $33\frac{1}{3}$ % of 54?

STEP 1   *question:* What is?

You are looking for the part.

STEP 2   *necessary information:* $33\frac{1}{3}$ %, of 54

$33\frac{1}{3}$ is the percent.

54 is the whole.

STEP 3   *proportion:*

$$\frac{n}{54} = \frac{33\frac{1}{3}}{100}$$

STEP 4   Cross multiply and divide.

$$100 \times n = 33\frac{1}{3} \times 54$$
$$100n = 1{,}800$$
$$n = \frac{1{,}800}{100} = \mathbf{18}$$

EXAMPLE 2   Bob takes home \$156.40 out of his weekly pay of \$184.00. What percent of his pay does he take home?

STEP 1   *question:* What percent of his pay does he take home?

You are looking for the percent.

STEP 2   *necessary information:* \$156.40, \$184.00

\$156.40 is the part.
\$184.00 is the whole.

STEP 3   *proportion:*

$$\frac{\$156.40}{\$184.00} = \frac{n}{100}$$

STEP 4   Cross multiply and divide.

$$184 \times n = 156.40 \times 100$$
$$184n = 15{,}640$$
$$n = \frac{15{,}640}{184} = \mathbf{85\%}$$

You can also use the percent circle to solve percent word problems with fractions or decimals.

EXAMPLE 3 Specialty Hardware needs to collect 6% sales tax on all items. In one day the store collected $424.80 in taxes. What were the store's total sales for the day?

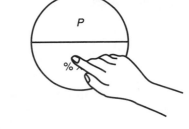

STEP 1 Since you are looking for the total sales for the day and you know the taxes collected and the tax rate, you are looking for the whole. Therefore, cover W (the whole) since that is the number you're trying to find.

STEP 2 Use the percent circle. $\frac{P}{\%}$ means $P \div \%$.

To find the whole, divide the part by the percent.

STEP 3 Convert the percent to a fraction or a decimal.

$6\% = \frac{6}{100}$ or 0.06

STEP 4 Do the calculation.

$$424.80 \div \frac{6}{100} = 424.80 \times \frac{100}{6} = \frac{42480}{6} = \$7,080, \text{ or } .06 \overline{)424.80}^{\;\$7,080}$$

The total sales for the day were **$7,080**.

···

Solve the following percent problems using proportions or the percent circle.

1. 4.5% of what number is 90?

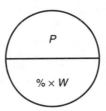

2. $\frac{1}{10}$ is what percent of $\frac{3}{4}$?

3. $66\frac{2}{3}\%$ of what number is 42?

4. What is 6.4% of 800?

5. Russo's Restaurant collected $49.76 in taxes Friday night. The food tax is 8%. How much money did the restaurant receive for meals on Friday night?

For problems 6–8, use the tax guidelines chart shown at the right.

State Tax Guidelines

Clothing—6%
Food—5%
Alcohol—7%
Cigarettes—8%

6. Jed bought a steak dinner for $8.60. Find the amount of tax he paid.

7. Sylvia bought a skirt for $42.50. Find the amount of tax she paid.

8. Flavia bought a bottle of wine costing $15.95 for a dinner party. Find the amount of tax she paid.

9. Glenda bought maple syrup for $1.92 a pint and sold the syrup for $0.96 a pint more. By what percent did she mark up the price of the maple syrup?

10. Barnes and Noble was having a storewide book sale in which all prices were cut $12\frac{1}{2}$%. How much did Juan save on a book that normally costs $12.80?

11. A telemarketing company found that 3.5% of its calls resulted in new subscribers. If its target is 12,000 new subscribers, how many calls should the company plan to make?

Solving Percent Word Problems

The following problems give you a chance to review percent word problems containing whole numbers, decimals, and fractions.

Solve the problems by using proportions or the percent circle.

1. Of the dentists surveyed, 3 out of 4 recommend a fluoride toothpaste. What percent of the dentists surveyed recommend a fluoride toothpaste?

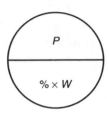

2. Of the registered voters, 112,492 people voted for mayor in the city. This was 40% of the registered voters. How many registered voters are there in the city?

3. In a normal season, the Seaside Resort has 34,500 visitors. This year, due to bad weather, 11,500 fewer visitors came to the resort. What was the percent drop in business for the resort?

4. The High Tech Electronics Company announced an 8.6% profit on sales of $49,600,000. How much profit did the company make?

5. The Quality Chocolate Company decided to increase the size of its chocolate bar 0.4 ounce. This was an increase in size of $16\frac{2}{3}$%. What was the weight of its chocolate bar before the change?

6. In a recent flu epidemic, 0.8% of people over age 65 who caught the disease died. The death toll in this group was 40. How many people over age 65 caught the flu?

7. In 2000, Dennis paid 13% of his income in taxes. How much did he pay in income taxes that year according to the chart at the right?

Yearly Earnings Dennis Ferguson	
1999	$10,048
2000	$11,694
2001	$12,509

8. The Machinist's Union has just won a 7% raise for its members. Dan is a union member who was making $17,548. How much of a raise will he get?

9. Nayana received a 9% raise worth $18 a week. What had her week's salary been?

10. Basketball star Kareem scored on 506 out of 1,012 attempts. What was his scoring percentage?

11. A $\frac{3}{4}$-cup serving of Honey Nut Cheerios served with skim milk provides 30% of the U.S. recommended daily allowance of vitamin A. How many cups of Honey Nut Cheerios should you eat in order to receive the full U.S. recommended daily allowance of vitamin A when you have skim milk with each serving?

12. A $3\frac{3}{4}$-ounce serving of Norway Sardines in chili sauce provides 100% of the U.S. recommended daily allowance for vitamin D. What percent of the U.S. daily allowance is provided per ounce of the Norway Sardines in chili sauce?

COMBINATION WORD PROBLEMS

Solving Combination Word Problems

Until now, this book has shown you one-step word problems. However, many situations require you to use a combination of math operations to solve word problems.

Generally, you can solve these combination problems by breaking them into two or more one-step problems. As you read a word problem, you may see that it will take more than one math operation to solve. The difficulty lies in deciding how many steps to take and in what order to work them out.

The key to solving combination problems is—

start with the question and work backward.

This shouldn't be difficult. Throughout this book, you have started your work with finding the question.

Follow these steps in solving combination word problems.

STEP 1 Find the question.

STEP 2 Select the necessary information.

STEP 3 Write a solution sentence for the problem. Fill in only the necessary information that belongs in the solution sentence.

Write another sentence, this time to find the information that is missing in the solution sentence. Solve the sentence that gives you the missing information.

STEP 4 Fill in the missing information (the answer from Step 3) in the solution sentence and solve.

STEP 5 Make sure that the answer is sensible.

No matter how many short problems a combination problem consists of, you can always work backward from the solution sentence. Examples 1 and 2 illustrate this.

EXAMPLE 1   Sengchen had $38 in her checking account. She wrote checks for $14 and $9. How much money was left in her checking account?

    STEP 1   *question:* How much money was left in her checking account?

    STEP 2   *necessary information:* $38, $14, $9

    STEP 3   Write a solution sentence.

         money − checks = money left

         Fill in the sentence with information that can be used to solve the problem.   *$38 − checks = money left*

         Decide what *missing information* is needed to solve the problem. Write a number sentence and solve.

         check + check = checks

         $14 + $9 = $\boxed{\$23}$ *$38 − $\boxed{\$23}$ = money left*

         Now you have the complete information needed to solve the problem.

    STEP 4   Solve.   *$38 − $23 = $15 left*

EXAMPLE 2   Lillie worked as a travel agent. She arranged a trip for 56 people at a cost of $165 for airfare plus $230 for hotel per person. How much money did she collect from the group?

    STEP 1   *question:* How much money did she collect?

    STEP 2   *necessary information:* 56 people, $165 airfare, $230 hotel accommodations

    STEP 3   Write a solution sentence.

         cost × number of people = total   *cost × 56 = total*

         Solve for *missing information.*

         airfare + hotel = cost

         $165 + $230 = $\boxed{\$395}$ *$\boxed{\$395}$ × 56 = total cost*

    STEP 4   Solve.   *$395 × 56 = $22,120 total*

The words that you use in the solution and missing information sentences may differ from what is presented here. What is important is that you break down the problem into smaller steps.

A combination word problem that needs both multiplication and division to be solved can often be written as one proportion instead of two separate word sentences.

EXAMPLE 3 Apples were sold at a cost of 58 cents for 2 pounds. How much did Michelle pay for 3 pounds of apples?

  STEP 1 *question:* How much did Michelle pay for 3 pounds of apples?

  STEP 2 *necessary information:* 2 pounds, 58 cents, 3 pounds

  STEP 3 Write a proportion to show the relationship between weight and cost.

  STEP 4 Fill in the appropriate numbers and solve.

$$\frac{pounds}{cents} = \frac{pounds}{cents}$$

$$\frac{2\ pounds}{58\ cents} = \frac{3\ pounds}{n}$$

$$2 \times n = 3 \times 58$$

$$n = \frac{174}{2} = 87\ cents$$

After some practice, you will be able to tell which problems can be solved with a proportion. In most cases, breaking down a problem into smaller problems with word sentences will be the best method for solution.

....

For each word problem, write two word sentences (a solution sentence and a missing information sentence) or a proportion. DO NOT SOLVE!

1. Tim earns $380 a week. Every week $149 in taxes and $16 in union dues are taken out of his paycheck. What is his take-home pay?

2. After starting the day with $41, Miguel spent $3 for lunch and $22 for gas. How much money did he have left by the end of the day?

3. A store bought 30 boxes of dolls for $720. If there were 8 dolls in a box, how much did each doll cost?

4. Samuel had $394 in his checking account. After he wrote a check for $187 and deposited $201, how much money was in his checking account?

5. Kelly bought five of the blouses shown at the right and one skirt. How much money did she spend on these clothes?

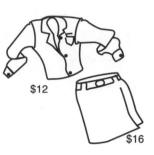

$12

$16

6. Martha borrowed $4,600 to buy a new car. She will have to pay $728 interest. She plans to pay back the loan plus the interest in 24 equal monthly payments. How much will her monthly payments be?

7. To be hired, a data entry operator must be able to enter numbers into a computer at the rate of 10,000 numbers every 60 minutes. Yvana took a 15-minute data entry test. How many numbers did she have to enter to be hired?

8. A 14-gram serving of Cain's Mayonnaise contains 5 grams of polyunsaturated fat and 2 grams of saturated fat. How many grams of saturated fat are there in a 224-gram jar of Cain's Mayonnaise?

9. At Burger Queen, a cheeseburger costs $1.89, medium french fries cost $1.29, and a medium cola costs $1.19. The cheeseburger combination meal of a cheeseburger, medium fries, and medium cola costs $4.29. How much does Olga save by buying the cheeseburger combination meal instead of buying the three items separately?

Write two one-step word sentences or a proportion needed to solve the following combination word problems. Then solve the problems.

10. For the convention, each of the 8 wards of the city elected 4 delegates, while 5 delegates were elected at large. How many delegates did the city send to the convention?

11. It cost the gas station owner $81 in parts and $45 in labor to fix his customer's car. He charged his customer $163. How much profit did the owner make on the job?

12. Mark was offered a job downtown that would give him a raise of $78 a month over his current salary, but his commuting costs would be $2 a day higher. If he works 22 days a month, what would be his net monthly increase in pay?

13. The recipe at the right serves 6 people. Ginny is planning to make the recipe for 8 people. How much stew beef does she need?

> ## Hearty Stew
>
> 3 cups beef stock
> 1 cup red wine
> $\frac{1}{4}$ cup oil
> 2 pounds potatoes
> 3 pounds stew beef
> 1 pound carrots

14. Last month Elvira was billed $60 for using 720 kilowatts of electricity. This month she checked her meter and found that she had used 648 kilowatts of electricity. Assuming the cost of electricity has not changed, what will her electric bill be for this month?

15. Carlos has a large landscaping job that requires 1,728 cubic feet of loam. His pickup truck can carry 48 cubic feet of loam. He can deliver 6 loads of loam each workday. How many workdays will it take Carlos to deliver all the loam he needs for his landscaping job?

Solving Problems with Decimals, Fractions, Percents

Decimal, fraction, and percent combination word problems are set up and solved in the same way as whole-number combination word problems.

EXAMPLE 1 For each child at her daughter's birthday party, Shelly spent $0.35 for a party favor and $0.16 for a balloon. She had 13 children at the party. How much did she spend for gifts for the children?

  STEP 1 *question:* How much did she spend for gifts for the children?

  STEP 2 *necessary information:* $0.35, $0.16, 13 children

  STEP 3 *solution sentences:*

  gifts × children = total cost

  missing information sentence:

  favor + balloon = gifts

  $0.35 + $0.16 = $0.51

  $$gifts \times 13 = total\ cost$$

  $$\$0.51 \times 13 = total\ cost$$
  $$\$0.51 \times 13 = \$6.63$$

  STEP 4 Solve.

EXAMPLE 2 Bright's department store advertised that everything in the store was $\frac{1}{5}$ off. Debbie bought a pair of pants labeled $20. How much did the pants cost her?

  STEP 1 *question:* How much did the pants cost her?

  STEP 2 *necessary information:* $\frac{1}{5}$, $20

  STEP 3 *solution sentence:*

  original price − discount = sale price

  missing information sentence:

  price × fraction = discount

  $20 × $\frac{1}{5}$ = $4 discount

  $$\$20 - discount = sale\ price$$

  $$\$20 - \$4 = sale\ price$$
  $$\$20 - \$4 = \$16$$

  STEP 4 Solve.

Both of these examples illustrate two-step word problems. Later in this chapter you will work with problems that need more than two steps for solution.

EXAMPLE 3 Real Value Hardware advertised that all prices had been reduced 15%. A socket set is on sale for $13.60. What was its original price?

    STEP 1 *question:* What was its original price?

    STEP 2 *necessary information:* 15%, $13.60

    STEP 3 *solution statement:* Since this is a percent problem, you can write a proportion.

$$\frac{part}{whole} = \frac{percent}{100} \qquad\qquad \frac{13.60}{n} = \frac{percent}{100}$$

                missing information:

                100% – percent reduced = percent sale

                100% – 15% = 85%

$$\frac{13.60}{n} = \frac{85}{100}$$
$$85 \times n = 13.60 \times 100$$
$$85n = 1,360$$
$$n = \$16$$

    STEP 4 Solve.

....

Write two one-step word sentences or a proportion needed to solve the following combination word problems. Then solve the problems.

1. Chris bought 6 boxes of cookies for $14.40. If there were 20 cookies in a box, how much did each cookie cost?

2. The $400 washing machine at the right was reduced for clearance. What was its sale price?

3. Beverly bought five cans of pears, each containing $9\frac{3}{4}$ ounces of pears, and one can of fruit cocktail containing $17\frac{1}{2}$ ounces of fruit. What was the total weight of the fruit she bought?

4. A water widget cost Phil $2.49. Because it reduced his use of hot water, it saved him $3.40 a month in costs for hot water. What were his net savings for 12 months?

5. When cooked, a hamburger loses $\frac{1}{3}$ of its original weight. How much does a $\frac{1}{4}$-pound hamburger weigh after it is cooked?

6. The tax on a meal is 6%. How much is Milton's total bill on a $24 dinner?

30% Off the Prices Below!

Shorts—regularly $10.50
Shirts—regularly $9.00
Socks—regularly $1.99
Bathing Suits—regularly $19.50

7. During the summer clearance sale, everything in the store was 30% off. Solaire bought the bathing suit advertised at the right. How much did she pay for the suit?

8. Marlene bought a new couch for $310.60. She paid $130 down and planned to pay the rest in 12 equal monthly payments. How much will she pay each month?

9. Walter bought a case of 30 bottles of cooking oil for $57.00. He then sold the oil for $2.10 per bottle. How much money did he make on each bottle?

10. Walter bought a case of 30 bottles of cooking oil for $57.00. He then sold the oil for a profit of $0.20 per bottle. What was the percent of profit? (Round to the nearest tenth of a percent.)

11. At her diner, Mireya added 0.2 ounce of salt to her 4 gallon (256 ounce) pot of beef stew. How much salt was in each 12-ounce portion?

12. A clothing manufacturer makes a blouse and matching skirt. The manufacturer buys material for the clothes in 60-inch-wide rolls. Each skirt pattern requires $\frac{2}{3}$ yard of material, while each blouse requires $1\frac{1}{4}$ yards of material. How many complete outfits can be made from a 70-yard-long roll of material?

Order of Operations

You've seen how to use solution sentences to solve combination word problems. If you know and use the rules for order of operations, you can write out the steps of a multistep word problem in one line.

Following are the rules for order of operations in arithmetic expressions.

Rule 1: Do multiplication and division before addition and subtraction.

EXAMPLE 1 $9 - 2 \times 4$

SOLUTION: Multiply. $2 \times 4 = 8$

Subtract. $9 - 8 = \mathbf{1}$

EXAMPLE 2 $24 \div 4 + 2$

SOLUTION: Divide. $24 \div 4 = 6$

Add. $6 + 2 = \mathbf{8}$

Rule 2: Do the arithmetic inside parentheses first.

EXAMPLE 1 $(21 - 6) \div 3$

SOLUTION: Subtract. $21 - 6 = 15$

Divide. $15 \div 3 = \mathbf{5}$

EXAMPLE 2 $5 \times (4 + 8)$

SOLUTION: Add. $4 + 8 = 12$

Multiply. $5 \times 12 = \mathbf{60}$

Rule 3: Using Rules 1 and 2, start solving arithmetic expressions from the left.

<u>EXAMPLE 1</u> $20 - 6 + 4$

        SOLUTION: Subtract. $20 - 6 = 14$

                   Add. $14 + 4 = \mathbf{18}$

<u>EXAMPLE 2</u> $1 + 2 \times 9 \div 6$

        SOLUTION: Multiply. $2 \times 9 = 18$

                   Divide. $18 \div 6 = 3$

                   Add. $1 + 3 = \mathbf{4}$

Using the rules for order of operations, evaluate the following arithmetic expressions.

1. $(5 \times 4) - (6 + 3) =$

2. $5 \times 4 - 6 + 3 =$

3. $145.6 + 12.2 - 5.7 - 1.1 =$

4. $3.2 \times 6 + 7.8 =$

5. $(7.8 + 2.2) \div 5 =$

6. $(4 \times 9) \div 3 + 6 =$

7. $4 \times 9 \div 3 + 6 =$

8. $1.8 \div 0.02 - 10 - 8 =$

9. $1.8 \div 0.02 - (10 - 8) =$

10. $56 - 7 \times 3.5 =$

Using Order of Operations in Combination Word Problems

Look at the following examples to see how to use your knowledge of order of operations to set up a multistep word problem.

EXAMPLE 1 Frank bought dinner for two for $22.50. His bill included a 5% meal tax. What was his total bill?

> **STEP 1** *question:* What was his total bill?
>
> **STEP 2** *solution sentence:*
>
> price of dinner + (5% × price of dinner) = total bill
>
> **STEP 3** *set up:*
>
> $22.50 + (0.05 × 22.50) = total bill
>
> **STEP 4** *solution:*
>
> Multiply. 0.05 × $22.50 = $1.125
>
> = $1.13 (nearest cent)
>
> Add. $22.50 + $1.13 = **$23.63**

EXAMPLE 2 Eblin lives 19 miles from work. Last week she drove to work and back 5 days and on the weekend drove 170 miles to visit relatives. How many miles did she drive last week commuting and visiting relatives?

> **STEP 1** *question:* How many miles did she drive?
>
> **STEP 2** *solution sentence:*
>
> commuting + visiting relatives = total miles
>
> **STEP 3** *set up:*
>
> (19 miles × 5 days × 2 times a day) + 170 miles = total miles
>
> **STEP 4** *solution:*
>
> Multiply. 19 × 5 × 2 = 190 miles
>
> Add. 190 miles + 170 miles = **360 miles**

For each word problem, circle the letter of the correct setup.

1. Three friends went out to dinner. Their meal cost $26.88. If they left a tip of $4.00, how much did each person pay if they divided the total amount equally?

   a. ($26.88 ÷ 3) + $4.00
   b. $26.88 + $4.00 ÷ 3
   c. $26.88 + ($4.00 × 3)
   d. ($26.88 + $4.00) ÷ 3
   e. ($26.88 + $4.00) × 3

2. Kathy and Peter went clothes shopping for their baby daughter, Gina. They bought five sleepers for $8.95 each and six T-shirts for $2.40 each. How much money did they spend?

   a. 5 × $8.95 + 6 × $2.40
   b. (5 + $8.95) × (6 + $2.40)
   c. (5 + 6) × ($8.95 + $2.40)
   d. ($8.95 ÷ 5) + ($2.40 × 6)
   e. ($8.95 + $2.40) ÷ (5 + 6)

3. Katrina pays $99.90 for her monthly train pass. If she used her pass twice a day for 23 days last month, what was the average cost of each ride?

   a. ($99.90 ÷ 23) × 2
   b. $99.90 ÷ (23 × 2)
   c. $99.90 − (23 × 2)
   d. $99.90 ÷ 23
   e. $99.90 ÷ 23 × 2

4. A coat normally selling for $80 was marked down 40%. What was the sale price?

   a. $80 − 40
   b. $80 ÷ 40
   c. $80 − 0.40 × $80
   d. $80 + 0.40 × $80
   e. $80 − $80 ÷ 40

5. Dolores needs to buy eggs for the week. She will make six 3-egg omelets for her family. She will also need four eggs for a cake and two eggs for a casserole. How many dozen eggs must she buy?

   a. (6 × 3) + 4 + 2
   b. (6 + 3 + 4 + 2) ÷ 12
   c. 12 × (6 ÷ 3) + (4 ÷ 2)
   d. 6 × 3 ÷ 12 + 4 + 2 ÷ 12
   e. (6 × 3 + 4 + 2) ÷ 12

Using Memory Keys for Multistep Word Problems

Now that you know the order of operations, you can use the memory keys on your calculator for multistep word problems.

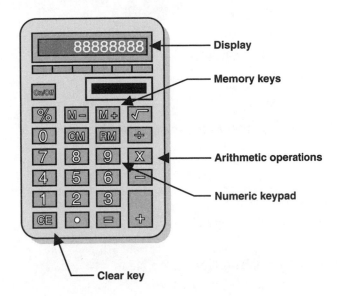

Display

Memory keys

Arithmetic operations

Numeric keypad

Clear key

First you need to write out your setup of the solution of a word problem. Then following the order of operations, do each calculation. You can either add or subtract the result to your calculator memory when you need to. The memory stores these results until it is cleared with either a Clear All or a Clear Memory.

There are some multistep calculations that do not need the memory keys at all, such as a series of multiplications. Practice with your calculator to see if you can find more than one correct procedure for solving a multistep word problem.

Following is an example of how you could use the memory keys to solve a multistep word problem. Since not all calculators handle memory the same way, work through this example on your calculator to make sure these directions work on it.

EXAMPLE   For a party, Hanan bought six bottles of soda for $0.89 each and three bags of chips for $1.49 each. How much did she spend?

STEP 1   Set up the solution.

cost of soda + cost of chips = total cost

$(6 \times 0.89) + (3 \times 1.49) = $ total cost

STEP 2   Key in the first operation in parentheses and add it to memory.

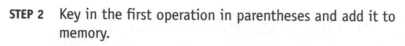

(You may need to use the = key before the memory key.)

You should see 5.34 in your display.

STEP 3   Key in the second operation in parentheses and add it to the amount already in memory.

$$\boxed{3}\;\boxed{\times}\;\boxed{1}\;\boxed{.}\;\boxed{4}\;\boxed{9}\;\boxed{M+}$$

(You may need to use the = key before the memory key.)

You should see 4.47 in your display.

STEP 4   Press $\boxed{RM}$ (sometimes labeled $\boxed{MRC}$ or $\boxed{RCM}$) to read the result. You should see 9.81 on the display.

STEP 5   Press $\boxed{CM}$ (Clear Memory) or $\boxed{CA}$ (Clear All) to clear the memory before you start calculating another problem.

....

 For each word problem, look at the solution setup and select the correct question. Then use the setup and your calculator to solve the problem.

1. Sugi has $46 left on her debit card and $48 in cash. She bought $81 in groceries.

    ($46 + $48) − $81

    a. How much more did Sugi have in cash?
    b. How much money did Sugi have left?
    c. How much money did she have before she paid for her groceries?
    d. If she had forgotten her debit card, how many additional dollars would she have needed to pay her bill?
    e. How much money does Sugi need to borrow to pay her bill?

2. Emily mixes a fruit cooler with 5 ounces of fruit juice to 3 ounces of seltzer. For a party she wants to make 2 gallons (164 ounces) of fruit cooler.

$$\frac{5}{5+3} = \frac{n}{256}$$

   a. How much seltzer will she need?
   b. How much more fruit juice than seltzer will she use?
   c. What is the size of one serving?
   d. How many servings is she making?
   e. How much fruit juice will she need?

3. The clearance advertisement said that an additional 20% would be taken off the labeled price at the register. Nerys bought a pair of boots marked $85 and a belt marked $25.

$$\frac{n}{\$85} = \frac{100-20}{100}$$

   a. How much did Nerys save on the boots?
   b. How much did Nerys pay for the boots?
   c. What percent of the original price did Nerys have to pay?
   d. How much did Nerys save on the two items?
   e. How much did Nerys save on the belt?

4. Ultra Clean detergent comes in two sizes. The 64-ounce size costs $8.68. The 32-ounce size costs $4.02.

$$(\$8.68 \div 64) - (\$4.02 \div 32)$$

   a. How much more does the larger size cost?
   b. How much more does the larger size cost per ounce?
   c. What is the per ounce cost of the larger size?
   d. What is the per ounce cost of the smaller size?
   e. How much more detergent is in the larger size?

Using Pictures or Diagrams to Set Up Multistep Word Problems

You can use pictures or diagrams to help you set up and solve multistep word problems.

EXAMPLE 1 Holly just moved into her new studio apartment. The main room is 30 feet long by 15 feet wide, and the bathroom is 9 feet long and 8 feet wide. What is the total size of her apartment in square feet?

STEP 1 *question:* What is the total size of her apartment?

STEP 2 *picture:*

STEP 3 *set up:*

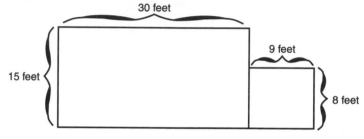

(30 feet × 15 feet) + (9 feet × 8 feet) = total size

STEP 4 *solution:*

Multiply. 30 feet × 15 feet = 450 square feet

Multiply. 9 feet × 8 feet = 72 square feet

Add. 450 square feet + 72 square feet = **522 square feet**

EXAMPLE 2 Donna's car gets 27 miles per gallon. After filling her 18-gallon gas tank, she drove 135 miles. How many more miles can she drive before her car runs out of gas?

STEP 1 *question:* How many more miles?

STEP 2 *diagram:*

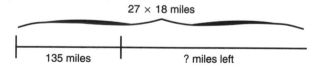

STEP 3 *set up:*

(27 mpg × 18 gallons) – 135 miles = miles left

STEP 4 *solution:*

Multiply. 27 mpg × 18 gallons = 486 miles

Subtract. 486 miles – 135 miles = **351 miles**

For each word problem, draw a picture or diagram and circle the letter of the correct setup.

1. A set of silverware included 8 place settings of a knife, fork, salad fork, teaspoon, and soup spoon as well as 6 serving pieces. How many pieces of silverware are in the set:

   a. 8×6 diagram:
   b. $8 \times 6 - 5$
   c. $8 \times 5 + 6$
   d. $5 + 8 \times 6$
   e. $8 \times 6 \div 5$

2. A florist has 420 roses and 380 carnations. How many bouquets, each containing 5 flowers, can be made from these flowers?

   a. $(420 + 380) \times 5$ diagram:
   b. $5 \times 420 - 380$
   c. $5 \div (420 + 380)$
   d. $420 \times 5 - 380 \times 5$
   e. $420 \div 5 + 380 \div 5$

3. Jan turned a 9-foot by 12-foot room in her home into an office. If she wants three equally sized bookcases along each of the 9-foot walls, how wide can each bookcase be?

   a. $(9 \div 3) \times 2$ diagram:
   b. $(9 \times 12) \div 3$
   c. $(9 \div 3) \times 12$
   d. $9 \div 3$
   e. $(12 \div 3) \times 9$

4. The theater has 30 rows of 25 seats each. If 150 of the seats are taken, how many are empty?

   a. $(30 \times 25) - 150$ diagram:
   b. $(30 \times 25) \div 150$
   c. $150 - 30 \times 25$
   d. $(30 + 25) \times 150$
   e. $(150 \div 30) + 25$

Solving Combination Word Problems Involving Conversions

Many combination word problems involve conversions. To solve the following examples and problems, you should refer to the conversion chart on page 219. You should notice that a conversion is needed when one unit of measurement appears in the necessary information and a different unit of measurement is called for in the question.

EXAMPLE 1 A dairy farm sold 156 *quarts* of milk at its own store and shipped out an additional 868 *quarts* to nearby supermarkets. How many *gallons* of milk were marketed?

STEP 1 *question:* How many gallons of milk were marketed?

STEP 2 *necessary information:* 156 quarts, 868 quarts

STEP 3 *solution statement:*

$$\frac{\text{quarts}}{\text{gallon}} = \frac{\text{quarts}}{\text{gallon}}$$

$$\frac{4}{1} = \frac{\text{quarts}}{\text{gallons}}$$

There are 4 quarts in a gallon. This is written on the left side of the proportion as $\frac{4}{1}$.

$$\frac{4}{1} = \frac{1{,}024}{n}$$

missing information sentence:

quarts + quarts = total quarts

156 + 868 = 1,024 quarts

$$4 \times n = 1 \times 1{,}024$$
$$4n = 1{,}024$$

STEP 4 Solve.

$$n = \frac{1{,}024}{4} = 256 \text{ gallons}$$

EXAMPLE 2 A mill is cutting 8-*foot* lengths of lumber into chair legs. There are 6 *inches* of scrap for each length. What percent of the wood is scrap?

STEP 1 *question:* What percent of the wood is scrap?

STEP 2 *necessary information:* 8-foot, 6 inches

STEP 3 *solution statement:*

$$\frac{\text{part}}{\text{whole}} = \frac{\text{percent}}{100}$$

$$\frac{6 \text{ inches}}{8 \text{ feet}} = \frac{n}{100}$$

Since all the information must be in the same unit of measurement, do the conversion.

conversion:

$$\frac{1 \text{ foot}}{12 \text{ inches}} = \frac{8 \text{ feet}}{x}$$

$$x = 12 \times 8$$

$$x = 96 \text{ inches}$$

> **Note:** In the conversion, the letter x was used to stand for the unknown number of inches. Any letter can be used to stand for an unknown.

STEP 4 Solve.

$$\frac{6 \text{ inches}}{96 \text{ inches}} = \frac{n}{100}$$

$$96n = 6 \times 100$$
$$96n = 600$$

$$n = \frac{600}{96} = 6\frac{1}{4}\%$$

 Solve the following word problems, making the necessary conversions. Be careful; not all problems need a conversion. Refer to the conversion chart on page 219 if necessary.

1. Tile Town sells 81-square-inch tiles. How many tiles are needed to cover a 54-square-foot floor?

2. On the airplane assembly line, Isadore was able to make 20 welds an hour. How many welds did he make during a 9-hour workday?

3. On Interstate Highway, there is a reflector every 528 feet. How many reflectors are there on the stretch of highway shown at the right?

    46 miles

    INTERSTATE HIGHWAY

4. The medical center needed 48 gallons of blood after the earthquake. A nearby city donated 26 gallons of blood. The rest was donated at the medical center by people each giving 1 pint of blood. How many people gave a pint of blood at the center?

5. The Heat Coal Company distributed 38 tons of coal to its customers in 1 day. It delivered 400 pounds of coal to each of its customers. How many customers received deliveries?

6. Sharon was able to type 463 numbers during a 5-minute timing for data entry. At this rate, how many numbers could she type in an hour?

7. Lynn brought 12 quarts of ice cream to the Fourth of July picnic. If she gives each person a 4-ounce serving of ice cream, how many people will get the ice cream?

Solving Word Problems Containing Unnecessary Information

Unnecessary information is more difficult to spot in combination word problems than in one-step word problems. The key to identifying unnecessary numbers is in working backward from the question. Once you write a word sentence, look at all the given information, and decide what is needed to answer the question.

EXAMPLE 1 At sunrise the temperature was 54 degrees. By midafternoon, it had risen 27 degrees. The temperature then began falling, until by midnight it had dropped 19 degrees from the high. What was the temperature at midafternoon?

STEP 1 *question:* What was the temperature at midafternoon?

STEP 2 *necessary information:* 54 degrees, 27 degrees (The fact that the temperature had dropped another 19 degrees by midnight is unnecessary information.)

STEP 3 This is a one-step problem. Write a word sentence.
sunrise temperature + change = midafternoon temperature

STEP 4 Solve.

$54 + 27 = $ **81 degrees**

EXAMPLE 2 A 0.8-ounce jar of basil sells for $0.98. Marie has 3.5 pounds of basil to pack into the jars. How many jars will she need?

STEP 1 *question:* How many jars will she need?

STEP 2 *necessary information:* 0.8 ounce, 3.5 pounds

(The cost of the jar of basil is unnecessary information.)

STEP 3 *solution statement:*

$$\frac{\text{total weight}}{\text{number of jars}} = \frac{\text{weight}}{1 \text{ jar}}$$

Since all your weights must be in the same unit of measurement, your next step must be a conversion to find the number of ounces in a pound.

conversion: $\dfrac{16 \text{ ounces}}{1 \text{ pound}} = \dfrac{n \text{ ounces}}{3.5 \text{ pounds}}$

$1 \times n = 3.5 \times 16$

$n = $ **56 ounces**

STEP 4 Solve.

$$\frac{3.5 \text{ pounds}}{n \text{ jars}} = \frac{0.8 \text{ ounce}}{1 \text{ jar}}$$

$$\frac{56 \text{ ounces}}{n \text{ jars}} = \frac{0.8 \text{ ounce}}{1 \text{ jar}}$$

$0.8 \times n = 56 \times 1$

$0.8n = 56$

$n = \dfrac{56}{0.8} = 70 \text{ jars}$

Write the word sentences or proportion needed to solve the following word problems. Underline the necessary information. Then solve the problem. Be careful; many of these problems contain unnecessary information.

1. At the beginning of the school year, the Philadelphia school system had 103,912 students. During the course of the year, 4,657 students left the system, while 1,288 more students were enrolled. What was the student population at the end of the year?

2. At the beginning of the school year, the Philadelphia school system had 103,912 students. During the course of the year, 4,657 students left the system, while 1,288 more students were enrolled. How many different students spent at least part of the year in the Philadelphia school system?

3. At sunrise the temperature was 54 degrees. By midafternoon, it had risen 27 degrees. The temperature then began falling, until by midnight it had dropped 19 degrees from the high. What was the temperature at midnight?

4. Every week, after having $153 taken out of his paycheck, Lloyd takes home $348. What was Lloyd's take-home pay for a 52-week year?

5. Ahmed bought three paperbacks and two magazines at the drugstore at the prices shown at the right. He paid for his purchases with a $50 bill. How much change did he receive at the drugstore?

6. Sangita is a member of a cooperative grocery store. She gets a 20% discount off everything she buys in the store. She bought a 5-pound bag of oranges marked $2.80. After receiving her discount, how much did she pay for the oranges?

Hoyle's Drugstore

Newspapers	$0.55
Paperbacks	$6.95
Magazines	$3.50
Postcards	$0.35

Solving Longer Combination Word Problems

Sometimes word problems cannot be solved by being broken into two one-step problems. Three or more steps may be needed to solve the problem. The method used with these problems is the same as the method used throughout this chapter with combination word problems. Keep working backward from the question. Set up a solution sentence and solve shorter problems to get all of the information that you need.

EXAMPLE   Sylvia went shopping in the bargain basement. She bought a $24.99 dress marked $\frac{1}{3}$ off and a $16.95 pair of pants marked down 20%. How much did she spend?

STEP 1   *question:* How much did she spend?

STEP 2   *necessary information:* $24.99, $\frac{1}{3}$ off; $16.95, marked down 20%

STEP 3   *solution sentence:*

dress price + pants price = total spent

To solve this, you must find the sale prices of both the dress and the pants. Both can be found by using this *missing information* sentence:

original price − discount amount = sale price

You can find the discount by multiplying the original amount by a fraction or percent.

dress

$24.99 − ($\frac{1}{3}$ × 24.99) = sale price
24.99 − 8.33 = $16.66

pants

$16.95 − (20% of 16.95) = sale price
16.95 − (.20 × 16.95) = sale price
16.95 − 3.39 = $13.56

STEP 4   Solve.

$16.66 + $13.56 = **$30.22**

> **Note:** An earlier chapter used the proportion method for solving percent word problems. However, if a problem requires you to find a percent of an amount, there is another method. Simply change the percent to a decimal and multiply. In the example above, 20% was changed to .20.

For every problem, write all necessary word sentences or proportions. Then solve the problem. Round money solutions to the nearest cent.

1. Kerry bought seven apples and a cantaloupe at the prices shown at the right. How much did she spend?

   ....
   : Cantaloupe $1.88 each   :
   :   :
   : Grapes $1.69 per pound   :
   :   :
   : Apples $2.76 per dozen   :
   ....

2. Aaron received a gas bill of $36.80 for 32 gallons of bottled gas. If he pays the bill within 10 days, he will receive a 6% discount. How much will he pay if he pays his bill within 10 days?

3. Harold's doctor advised him to cut down on the calories he consumes by 28%. Harold has been consuming 4,200 calories a day. If Harold's breakfast contains 797 calories, how many calories can he have during the rest of the day?

4. Sarkis is a salesman. He receives a salary of $70 a week plus a 6% commission on all his sales over $200. Last week he sold $4,160 worth of merchandise. What was he paid for the week?

5. Dinora drove 3,627 miles from coast to coast. Her car averaged 31 miles per gallon, and she spent $186 for gas. On the average, what did she pay per gallon of gas?

Solving Combination Word Problems

 In the following problems, choose the one best answer. Round decimals to the nearest cent or the nearest hundredth.

1. Every day Kevin has to drive 7 miles each way to work and back. At work, he has to drive his truck on a 296-mile delivery route. How many miles does he drive during a 5-day workweek?

   a. 310 miles
   b. 315 miles
   c. 1,550 miles
   d. 4,214 miles
   e. none of the above

2. Every day, Jason has a 14-mile round-trip drive to work. He then has to drive his truck on a 296-mile delivery route 5 days a week. How many miles does he drive each day?

   a. 310 miles
   b. 315 miles
   c. 1,550 miles
   d. 4,214 miles
   e. none of the above

3. For his art class, Karl spent $135 on books and $225 on materials. To cover costs, how much did each of his 15 students pay?

   a. $360
   b. $90
   c. $24
   d. $15
   e. $9

4. Jessie's restaurant had four small dining rooms with a capacity of 28 people each and a main dining room with a capacity of 94 people. What was the total capacity of the restaurant?

   a. 126 people
   b. 658 people
   c. 348 people
   d. 122 people
   e. 206 people

5. Each team in the 8-team football league used to have a roster of 36 players. The league decided to decrease each team's roster size by 3 players. Before the change, how many players were in the league?

   a. 180 players
   b. 288 players
   c. 285 players
   d. 264 players
   e. 396 players

6. Each team in the 8-team football league used to have a roster of 36 players. The league decided to decrease each team's roster size by 3 players. After the change, how many players were in the league?

   a. 180 players
   b. 288 players
   c. 285 players
   d. 264 players
   e. 396 players

7. At the supermarket, Monique bought 2.36 pounds of cheese and 4 pounds of apples. What was the total cost of the cheese and apples at the prices shown at the right?

   a. $6.36
   b. $6.09
   c. $3.56
   d. $9.65
   e. $9.44

   **Meg's Market
   On Sale This Week!**

   Apples—only $0.89/pound
   Chicken—only $1.39/pound
   Potato Salad—only $1.29/pint
   Cheese—only $2.58/pound

8. A bottle contains 6 cups of laundry detergent. The directions say to use $\frac{1}{3}$ cup for a top-loading washer and $\frac{1}{4}$ cup for a front-loading washer. How many more loads per bottle can you do with a front-loading washer than with a top-loading washer?

   a. 1 load
   b. 3 loads
   c. 6 loads
   d. 8 loads
   e. 9 loads

Word Problems Posttest A

This posttest gives you a chance to check your skill at solving word problems. Take your time and work each problem carefully. When you finish, check your answers and review any topics on which you need more work. (**Caution:** At least one problem does not contain enough information to solve the problem.)

1. Three tablespoons cocoa plus 1 tablespoon fat can be substituted for 1 ounce chocolate in baking recipes. A recipe for chocolate cake calls for 12 ounces of chocolate. If Shirley is substituting cocoa for chocolate, how much cocoa should she use?

2. Matt has a 400-square-inch board. He needs a 25-square-inch piece of the board for the floor of a birdhouse. What percent of the board will he need for the floor of the birdhouse?

3. There are 5,372 school-age children in town. Of those children, 1,547 either go to private school or have dropped out. How many children remain in the town's public schools?

4. A bushel of apples weighs 48 pounds. Tanya wants to buy 12 pounds of apples. How many bushels should she buy?

5. At the sidewalk stand, Jason bought a hot dog and a soda. How much did he spend at the prices shown at the right?

HOT DOG	1.30
ITALIAN SAUSAGE	1.95
POTATO CHIPS	.50
SODA	.65

HOT DOGS

6. If Kenneth retires at age 65, he will receive as a pension 80% of his salary of $28,657. If he retires at age 62, he will receive only 70% of his salary. How much less will he receive for his pension if he retires early?

7. One cup sugar plus $\frac{1}{4}$ cup liquid can be substituted for 1 cup corn syrup in baking recipes. A recipe calls for $1\frac{1}{2}$ cups corn syrup. If Mira is substituting sugar for corn syrup, how much liquid should she add?

8. A conservation organization charged each member $20 dues plus $15 for the organization's magazine. How much money did the organization collect from its 13,819 members?

9. Pat and Connie are able to put down $16,000 as a down payment on a new home. Their bank told them that they must pay at least 8% of the purchase price as a down payment. What is the most expensive home they can afford?

10. For her wardrobe, Mrs. Are was given a Paris original worth $1,346, a New York original worth $658, and a Goodwill original worth $4. What was the total value of the clothes given to her?

11. After 3 years, Elsie's car had lost $\frac{1}{3}$ of its original value. Two years later, it had lost an additional $\frac{1}{4}$ of its original value. If she bought the car for $9,600, what was its value after 5 years?

12. There are 3 feet in a yard. There are 1,760 yards in a mile. Connie is planning to walk the 5-mile Walk for Peace. How many feet long is the Walk for Peace?

13. Lorraine weighed $172\frac{1}{2}$ pounds. She lost $47\frac{3}{4}$ pounds in one year. What was her new weight?

14. To qualify for the car race, Christine needed to drive 100 miles in under 43 minutes. She completed the first lap in $4\frac{1}{2}$ minutes. At this rate, what will be her total time for the 100-mile qualifying distance?

15. Nickilena, Jean, Rosemary, and Elaine went into business together. The four-woman partnership earned $336,460 and had expenses of $123,188. If they divided the profits equally, how much did each woman make?

16. Sears is offering 20% off on its $260 refrigerator. How much can you save by buying the refrigerator on sale?

17. Before a recent election, $\frac{1}{3}$ of the voters polled said they were planning to vote for the incumbent, while $\frac{1}{4}$ said they were planning to vote for the challenger. The rest were undecided. What fraction of the voters had decided which way they were going to vote?

18. Denise is paid $7.20 per hour at her part-time job. Last week she worked 17.5 hours. How much did she earn last week at her part-time job?

19. Jane bought a new hatchback automobile. She ordered $4,350 of added options and received a $2,650 rebate. How much did Jane pay for the car?

20. East Somerville has 948 homes. The Heart Association has 12 people collecting donations. If they all visit the same number of homes, how many homes should each of them visit?

21. Melvin received an electric bill for $86.29. He knows that it cost him $59.00 a month for his air conditioning. How much would his bill have been if he had not operated the air conditioner?

22. Eileen bought three pairs of socks for $1.79 each and four towels for $2.69 each. How much did she spend?

23. Steve's Ice Cream Store puts $\frac{1}{16}$ pound of whipped cream on every sundae. For how many sundaes will the container of whipped cream pictured at the right last?

24. After paying $24.43 for dinner and $7.50 for a movie, Florence paid the baby-sitter $20.00. How much did the evening cost her?

25. A pile of books weighed 34.2 pounds. If each book weighed 0.6 pound, how many books were in the pile?

Word Problems Posttest A Prescriptions

Circle the number of any problem that you miss. A passing score is 20 correct answers. If you passed the test, go on to Using Number Power. If you did not pass the test, review the chapters in this book or refer to these practice pages in other materials from Contemporary Books.

PROBLEM NUMBERS	PRESCRIPTION MATERIALS	PRACTICE PAGES
1, 3, 10, 20	whole numbers	18–39, 61–76
	Math Exercises: Whole Numbers and Money	3–29
	Real Numbers: Estimation 1	5–36
	Breakthroughs in Math: Book 1	7–133
4, 7, 13, 17, 23	fractions	52–60, 80–88
	Math Exercises: Fractions	3–29
	Real Numbers: Estimation 2	1–38
	Math Skills That Work: Book 2	70–105
	Breakthroughs in Math: Book 2	66–113
5, 18, 21, 24, 25	decimals	40–51, 56–60, 77–79, 87–88
	Math Exercises: Decimals	3–29
	Real Numbers: Estimation 1	37–64
	Math Skills That Work: Book 2	32–69
	Breakthroughs in Math: Book 2	34–65
2, 9, 16	percents	118–136
	Math Exercises: Percents	3–29
	Real Numbers: Estimation 2	39–64
	Math Skills That Work: Book 2	106–139
	Breakthroughs in Math: Book 2	114–141
12	conversion	100–101, 154–155
	Math Exercises: Measurement	6–24
	Breakthroughs in Math: Book 1	134–152
14, 19	not enough information	110–114
	Math Exercises: Problem Solving and Applications	18–23
6, 8, 11, 15, 22	multistep word problems	137–161
	Math Exercises: Problem Solving and Applications	3–16

For further word problems practice:

Math Solutions (software)
    Whole Numbers; Fractions; Decimals;
    Percents, Ratios, Proportions
Pre-GED/Basic Skills Interactive (software)
    Mathematics Units 1, 2, 3, 6, 7
GED Interactive (software)
    Mathematics Units 2, 3, 4, 5

Word Problems Posttest B

This test has a multiple-choice format much like the GED and other standardized tests. Take your time and work each problem carefully. Round decimals to the nearest cent or the nearest hundredth. Circle the correct answer to each problem. When you finish, check your answers at the back of the book.

1. Large eggs weigh $1\frac{1}{2}$ pounds per dozen. Dawn bought eight large eggs. How much did the eggs weigh?

   a. 3 ounces
   b. $\frac{1}{3}$ pound
   c. 1 pound
   d. 18 ounces
   e. not enough information given

2. Glenn, the owner of a hardware store, originally paid $540.60 for 15 tool sets. At his year-end clearance sale, he sold the last tool set for $24.00. How much money did he lose on the last tool set?

   a. $180.60
   b. $1.50
   c. $12.04
   d. $36.04
   e. none of the above

3. Manny was working as a hot dog vendor. He sold a total of 426 hot dogs in one weekend. If he sold 198 on Saturday, how many did he sell on Sunday?

   a. 624 hot dogs
   b. 332 hot dogs
   c. 228 hot dogs
   d. 514 hot dogs
   e. none of the above

4. The television announcer reported that Elizabeth Quezada had received 39% of the votes in the race for mayor. The totals board behind the announcer showed that Elizabeth had received 156,000 votes. How many votes were cast in the election?

   a. 40,000 votes
   b. 400,000 votes
   c. 156,039 votes
   d. 608,480 votes
   e. 60,840 votes

5. On the average, Morriston Airport has 96 jumbo jets arriving each day. Each jumbo jet has an average of 214 passengers. How many passengers arrive by jumbo jet at Kennedy Airport each day?

    a. 310 passengers
    b. 20,544 passengers
    c. 118 passengers
    d. 222 passengers
    e. none of the above

6. To finish off the room, Ed needs a tile only $\frac{1}{3}$ foot wide. How much did he have to cut off the tile pictured at the right so that it would fit?

    a. $\frac{1}{2}$ foot
    b. $\frac{5}{12}$ foot
    c. $\frac{4}{7}$ foot
    d. $\frac{1}{4}$ foot
    e. $\frac{4}{9}$ foot

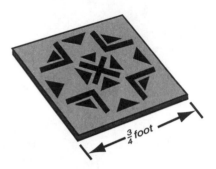

$\frac{3}{4}$ foot

7. After picking a bushel of apples, Tina planned to divide the apples equally among herself and five of her neighbors. How many apples did each of them get if the bushel weighed 54 pounds?

    a. 9 apples
    b. 60 apples
    c. 48 apples
    d. 324 apples
    e. not enough information given

8. Monty uses 1.23 cubic yards of concrete to cover 100 square feet with 4 inches of concrete. How many cubic yards does he need to cover 550 square feet with 4 inches of concrete?

    a. 650 square feet
    b. 4.92 cubic yards
    c. 6.77 cubic yards
    d. 2,200 square inches
    e. 27.06 cubic yards

9. Pedro needs 4 pounds of hamburger for his chili recipe. In his freezer, he has a 2.64-pound package of hamburger. How much more hamburger does he need?

   a. 6.64 pounds
   b. 2.68 pounds
   c. 1.36 pounds
   d. 1.52 pounds
   e. 10.56 pounds

10. Last year the city's supermarkets sold 1,638,000 gallons of milk. There are 78,000 people in the city. On the average, how many gallons of milk did each person buy?

    a. 1,716,000 gallons
    b. 1,560,000 gallons
    c. 21 gallons
    d. 127,716 million gallons
    e. not enough information given

11. After having $98.23 taken out of his paycheck, Maurice takes home $332.77 every week. What are Maurice's total gross earnings for a 52-week year?

    a. $12,196.08
    b. $22,412.00
    c. $431.00
    d. $17,304.04
    e. $5,107.96

12. In the last year 423 service stations in the state closed. Only 2,135 remain. How many service stations existed in the state a year ago?

    a. 2,558 service stations
    b. 1,712 service stations
    c. 2,312 service stations
    d. 2,512 service stations
    e. none of the above

13. A two-thirds majority of those voting in the House of Representatives is needed to override a presidential veto. If all 435 representatives vote, how many votes are needed to override a veto?

   a. 290 votes
   b. 145 votes
   c. 657 votes
   d. 658 votes
   e. 224 votes

14. Naomi had $61 in her checking account. She wrote a check for $28 and made a deposit. How much money did she then have in the account?

   a. $89
   b. $33
   c. $2.18
   d. $1708
   e. not enough information given

15. Avi needed to replace the molding on the left side of his car after it was damaged. The door needed $33\frac{3}{4}$ inches of molding, while the rear quarter panel needed $51\frac{2}{3}$ inches. How many inches of molding did he need if he replaced the molding on the door and the rear quarter panel?

   a. $2{,}247\frac{5}{12}$ inches
   b. $85\frac{5}{12}$ inches
   c. $17\frac{11}{12}$ inches
   d. $\frac{87}{124}$ inches
   e. $\frac{124}{87}$ inches

16. Ben, who works at the meat counter at the local supermarket, had to price the meat yesterday because the machine that normally did the job was broken. What price should he put on a 2.64-pound rib roast selling at $3.96 a pound?

   a. $1.50
   b. $5.94
   c. $6.60
   d. $10.45
   e. $1.32

17. At the factory, 28% of the workers were women. There were 432 male workers. What was the total number of workers at the factory?

    a. 460 workers
    b. 12,096 workers
    c. 600 workers
    d. 1,543 workers
    e. 404 workers

18. Sarah bought the carton of nails pictured at the right. How much did each nail weigh?

    a. 56 pounds
    b. 100 pounds
    c. 0.01 pound
    d. $\frac{1}{56}$ pound
    e. $\frac{3}{100}$ pound

19. In the first quarter, the Philadelphia 76ers hit only 7 out of 25 field goal attempts. What was their scoring percentage?

    a. 28%
    b. 32%
    c. 72%
    d. 76%
    e. 18%

20. Carla gained 3 pounds in the first month of her new diet and 4 pounds in the second month. Her original weight was 104 pounds. What was her new weight?

    a. 97 pounds
    b. 105 pounds
    c. 103 pounds
    d. 111 pounds
    e. 100 pounds

21. Peg bought the roast and the steak shown at the right. How much meat did she buy?

    a. 4.92 pounds
    b. 3 pounds
    c. 4.54 pounds
    d. 2.46 pounds
    e. 0.33 pound

1.23 pound steak

3.69 pound roast

22. Glennie had $74.81 in her checking account. She wrote checks for $46.19 and $22.45. She then made a $60.00 deposit. What was her new balance?

    a. $203.45
    b. $66.17
    c. $83.45
    d. $53.83
    e. $38.55

 23. Premium Ice Cream is 9% milkfat. How many pounds of milkfat are in a 450-pound batch of Premium Ice Cream?

    a. 50 pounds
    b. 441 pounds
    c. 459 pounds
    d. 40.5 pounds
    e. 2 pounds

 24. At the start of a trip, Tony filled his gas tank. After driving 168 miles, he needed 5.6 gallons of gasoline to fill his tank. How many gallons of gasoline would he use for the 417-mile drive from his home to his brother's home?

    a. 44.5 gallons
    b. 13.9 gallons
    c. 74.5 gallons
    d. 8.3 gallons
    e. 19.5 gallons

25. A piece of cheese was labeled $1.79 a pound. The price of the cheese was $1.06. How much did the cheese weigh?

    a. $2.85
    b. $0.73
    c. 1.69 pounds
    d. 1.90 pounds
    e. 0.59 pound

Word Problems Posttest B Chart

If you missed more than one problem on any group below, review the practice pages for those problems. Then redo the problems you got wrong before going on to Using Number Power. If you had a passing score, redo any problem you missed and go on to Using Number Power on page 175.

PROBLEM NUMBERS	SKILL AREA	PRACTICE PAGES
3, 12, 20	add or subtract whole numbers	18–39
5, 10	multiply or divide whole numbers	61–76
6, 15	add or subtract fractions	52–60
13, 18	multiply or divide fractions	80–88
9, 21	add or subtract decimals	40–51, 56–60
16, 25	multiply or divide decimals	77–79, 87–88
4, 19, 23	percents	118–136
1	conversion	100–101, 154–155
7, 14	not enough information	110–114
2, 8, 11, 17, 22, 24	multistep word problems	137–161

Using
Number
Power

Using Information from a Chart

Even a relatively small chart can contain a great amount of useful information. Sometimes you will need to use your problem-solving and math skills to understand and utilize the information contained in a chart.

Chart of Beverages

Serving Size	Beverage	Calories	Carbohydrates	Protein	Fat
8 ounces	2% lowfat milk	113 cal	11 grams	7 grams	4 grams
12 ounces	vanilla milk shake	381 cal	60 grams	13 grams	10 grams
8 ounces	orange juice from concentrate	91 cal	22 grams	1 gram	0 gram
12 ounces	cola	133 cal	34 grams	0 gram	0 gram

Georgiana wanted to gain weight. How many more calories would she get from a 12-ounce serving of the highest calorie beverage compared to a 12-ounce serving of the next highest calorie beverage?

At first glance, you might think that the cola is the next highest calorie beverage, but the serving size is larger than the milk or the orange juice. Since the milk and the orange juice are both 8-ounce servings and the milk is higher in calories, you can eliminate the juice. You can now set up a proportion to find the calories in a 12-ounce serving of low-fat milk.

$$\frac{8 \text{ ounces}}{12 \text{ ounces}} = \frac{113 \text{ calories}}{n \text{ calories}}$$

$8n = 1{,}356$

$n = \textbf{169.5 calories}$ in 12 ounces of milk

So 12 ounces of milk is the next highest calorie beverage.

Highest calorie beverage – next highest calorie beverage
= difference in calories

381 calories – 169.5 calories = **211.5 calories**

Use the chart of beverages to solve the following word problems.

1. For breakfast Danielle had an 8-ounce glass of orange juice. At lunch she had a 12-ounce milk shake, and for dinner she drank a 12-ounce bottle of cola. How many grams of carbohydrates did she get from beverages at the three meals?

   a. 32 grams
   b. 38 grams
   c. 116 grams
   d. 127 grams
   e. 505 grams

2. Keion had an 18-ounce vanilla milk shake. How many grams of fat were in his milk shake?

   a. 10 ounces
   b. 12 ounces
   c. 13 ounces
   d. 15 ounces
   e. 16 ounces

Use the nutrition chart to solve the following word problems.

	Calories	Carbohydrates	Protein	Fat
12-ounce sirloin steak	914	0 g	94 g	57 g
5-ounce hamburger on bun	405	24 g	27 g	21 g
6-ounce fried chicken breast	442	15 g	42 g	22 g
10-ounce spaghetti with tomato sauce	246	52 g	8 g	1 g
8-ounce broiled salmon	412	0 g	61 g	17 g

3. When Jean looked at the nutrition chart, she had trouble comparing the different main dishes because the portions were different sizes. She was trying to decide whether to make a 12-ounce portion of sirloin steak or a 12-ounce portion of broiled salmon. How many more calories was the steak than the salmon?

   a. 914 calories
   b. 1,326 calories
   c. 502 calories
   d. 296 calories
   e. 1,532 calories

4. Elba wanted to prepare a main dish that was approximately
   600 calories. If she prepared one of the main dishes from the chart,
   how large should a portion of that main dish be?

5. Max was going to have a 10-ounce portion of sirloin steak, but then
   decided he needed to cut down on the amount of fat in his diet.
   How much less fat would he eat if he instead had a 10-ounce
   portion of one of the other main dishes on the chart?

6. Using the chart as a guide, where does the grams of carbohydrates
   in a hamburger on bun come from, the hamburger or the bun or
   both?

7. Emily decided she needed more protein in her diet. She also wanted
   to have less fat in her diet. Which entrée would be the best choice
   for her?

   a. sirloin steak
   b. hamburger on bun
   c. fried chicken breast
   d. spaghetti with tomato sauce
   e. broiled salmon

Use the exercise chart to answer the following questions.

Exercise for a 130-Pound Woman	
Walking 3 mph	5 calories/minute
Jogging 5.5 mph	10 calories/minute
Running 10 mph	19 calories/minute
Swimming 25 yd/min	4 calories/minute
Bicycling 6 mph	4 calories/minute
Tennis singles	6 calories/minute
Cross-country skiing	6 calories/minute
Aerobic dancing	8 calories/minute
Recreational volleyball	3 calories/minute
Sleeping	1 calorie/minute

Match questions 8 through 10 to the correct solutions listed below.
Some questions have more than one solution.

- **a.** recreational volleyball for 90 minutes + tennis for 55 minutes
- **b.** running for 25 minutes + walking for 45 minutes + bicycling for 25 minutes
- **c.** jogging for 40 minutes + swimming for 30 minutes + bicycling for 2 hours
- **d.** aerobic dancing for 40 minutes + tennis for 40 minutes + jogging for 24 minutes
- **e.** cross-country skiing for 2 hours + aerobic dancing for 35 minutes

8. How could Fernande burn 1,000 calories? _____

9. Rafaela wanted to spend no more than 2 hours doing at least three different activities burning a total of 800 calories. How could she do it? _____

10. Juliette wanted to burn 600 calories while exercising at least 2 hours. How could she do it? _____

11. Design a fitness program for yourself using the numbers from the table (even if you are not a 130-pound woman). Plan to burn exactly 500 calories in at least an hour of activity.

12. Design an exercise program that is exactly one hour long and consists of at least two different activities. How many calories would a 130-pound woman burn if she followed your exercise program?

Challenging Word Problems

Sometimes word problems require more than translating words into mathematics. You might have to do more complex problem solving.

EXAMPLE Juana's recipe for chili calls for 2 teaspoons of hot sauce per gallon. Unfortunately, she misread the instructions and put in 2 tablespoons of hot sauce. How much more chili should she make in order to readjust her chili back to the recipe?

STEP 1 *question:* How much more chili should she make in order to readjust her chili back to the recipe?

STEP 2 *necessary information:* 2 teaspoons per gallon, 2 tablespoons

STEP 3 Decide what arithmetic operation to use. Convert tablespoons to teaspoons. Then use a proportion to find the total amount of chili.

Final equation: total amount of chili − chili already made = additional chili

STEP 4 There are 3 teaspoons in a tablespoon.

$$\frac{3 \text{ teaspoons}}{1 \text{ tablespoon}} = \frac{n \text{ teaspoons}}{2 \text{ tablespoons}}$$

$$n = 6 \text{ teaspoons}$$

STEP 5 Calculate the new total amount of chili.

$$\frac{2 \text{ teaspoons}}{1 \text{ gallon}} = \frac{6 \text{ teaspoons}}{x \text{ gallons}}$$

$$2x = 6$$

$$x = 3 \text{ gallons}$$

STEP 6 Calculate how much more chili needs to be made.

1 gallon chili + g gallons to be made = 3 gallons of chili

$g = 2$ gallons of chili

STEP 7 Does the answer make sense?

If three times the original amount of hot sauce is added to the recipe, you should end up with three times the original amount of chili. You will have correctly readjusted the recipe.

Solve the following word problems.

1. After a large storm, Pablo had 8 inches of water in his 30-foot by 40-foot basement. His sump pump pumped out 200 cubic feet of water an hour. Assuming no more water seeped into the basement, how long would it take to pump all the water out of the basement?

2. The same storm left 9 inches of water in Chantel's 30-foot by 50-foot basement, and some water continued to seep in for 4 hours. Her sump pump pumped out 300 cubic feet of water an hour. It took 6 hours to pump the water out of her basement. How much water seeped into the basement?

3. Awilda needs to get gas for her car, buy groceries, shop for a blouse at the department store, take out a book from library, and return a video to the video store. According to the map, what is the shortest total distance she will have to drive to do all her chores and return home?

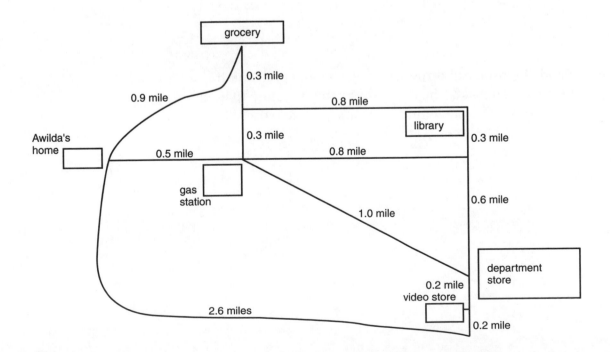

4. It takes 5 seconds for the sound of thunder to travel one mile. Light travels so fast that we can assume we are seeing lightning instantly. A storm is rapidly approaching. Domingo sees a lightning flash and hears the thunder 15 seconds later. Ten minutes later he sees another lightning flash and hears thunder 10 seconds later. If the storm continues to move at the same rate, how long will it take before it is directly overhead?

5. John is designing a 400-square-foot garden. He wants the garden to have the smallest possible perimeter, so he can spend as little as possible on fencing. What should the dimensions of his garden be?

6. Cammy has 100 feet of fencing for a running area for her dog. The run has to be at least 3 feet wide. How long could the run be?

7. Pitcher Randy Johnson can throw a fastball 95 miles per hour. It is 60.5 feet from the rubber at the top of the pitcher's mound to home plate. How long does it take Randy Johnson's fastball to travel from the pitcher's mound to home plate?

Answer Key

Pages 1–7, Pretest

1. b. weight per Chevrolet × number of Chevrolets = total weight
1,600 pounds × 840 Chevrolets = **1,344,000 pounds**

2. b. current height – growth = height at beginning of year
$48\frac{3}{8}$ inches – $2\frac{1}{4}$ inches = $46\frac{1}{8}$ **inches**

3. b. total cost per girl × number of girls = total collected
cost to get in + cost of roller blades = cost per girl
.75 + .50 = 1.25
1.25 × 19 girls = **$23.75**

4. a. $\frac{\text{part}}{\text{whole}} = \frac{\text{percent}}{100}$
$\frac{\$79.50}{\$n} = \frac{75\%}{100}$
75 × n = 79.50 × 100
n = 7,950 ÷ 75 = **$106**

5. d. $\frac{\text{concentrate}}{\text{water}} = \frac{\text{concentrate}}{\text{water}}$
$\frac{5 \text{ tablespoons}}{2 \text{ cups}} = \frac{n \text{ tablespoons}}{24 \text{ cups}}$
2 × n = 5 × 24
n = 120 ÷ 2 = **60 tablespoons**

6. c. $\frac{\text{part}}{\text{whole}} = \frac{\text{percent}}{100}$
$\frac{\$684}{\$3,600} = \frac{n\%}{100}$
3,600 × n = 684 × 100
n = 68,400 ÷ 3,600 = **19%**

7. c. turkey + chicken = meat
10.34 + 5.17 = **$15.51**

8. e. $\frac{200,000 \text{ pellets}}{n \text{ boxes}} = \frac{400 \text{ pellets}}{1 \text{ box}}$
400 × n = 1 × 200,000
n = 200,000 ÷ 400 = **500 boxes**

9. d. *conversion: 1 dozen = 12 oranges*
$\frac{\$2.40}{12 \text{ oranges}} = \frac{\$n}{4 \text{ oranges}}$
12 × n = 2.40 × 4
n = 9.60 ÷ 12 = **$0.80**

10. d. number of pounds ÷ number of slices = weight of each slice
3.15 pounds ÷ 24 slices = **0.13 pound**

11. e. You need to know how many inches are on a roll of masking tape.

12. a. girder weight × number of girders = total weight
$\frac{7}{8}$ ton × 600 = **525 tons**

13. b. sale price + reduction = original price
190 + 98 = **$288**

14. d. $\frac{n \text{ undecided}}{\text{total people}} = \frac{\text{percent undecided}}{100}$
100% – (in favor + against) = percent undecided
100% – (68% + 25%) = 100% – 93% = 7% undecided
$\frac{n}{1,400 \text{ people}} = \frac{7}{100}$
100 × n = 1,400 × 7
n = 9,800 ÷ 100 = **98 people**

15. b. $\frac{1\frac{1}{4} \text{ pounds}}{\$7.80} = \frac{1 \text{ pound}}{\$n}$
$\frac{5}{4} \times n = 7.80 \times 1$
$n = 7.80 \times \frac{4}{5} = $ **$6.24**

16. c. gas tank – new gas = old gas
18 gallons – 12.78 gallons = **5.22 gallons**

17. b. original price – reduction = sale price
fraction off × original price = reduction
$\frac{1}{3} \times 96 = 32$
96 – 32 = **$64**

18. a. current rent – increase = old rent
780 – 65 = **$715**

19. a. total tablets – tablets taken = tablets left
tablets per day × number of days = tablets taken
4 × 30 = 120
250 tablets – 120 tablets = **130 tablets**

20. c. *conversion: 16 ounces = 1 pound*
$\frac{2 \text{ ounces}}{1 \text{ box}} = \frac{n \text{ ounces}}{1000 \text{ boxes}}$
1 × n = 1,000 × 2
n = 2,000 ounces = 2,000 ÷ 16 = **125 pounds**

21. a. $\frac{\text{part}}{\text{whole}} = \frac{\text{percent}}{100}$
$\frac{n \text{ questionnaires}}{6,000 \text{ questionnaires}} = \frac{15\%}{100}$
100 × n = 15 × 6,000
n = 90,000 ÷ 100 = **900 questionnaires**

22. d. original gallons − gallons delivered = gallons left
deliveries × gallons per delivery = gallons delivered
7 deliveries × 364 gallons = 2,548 gallons delivered
9,008 − 2,548 = **6,460 gallons left**

23. d. new price + drop in value = old price
$8\frac{3}{4}$ dollars + $1\frac{7}{8}$ dollars = **$10\frac{5}{8}$ dollars**

24. e. number of pounds × price per pound = total cost
1.62 pounds × 2.43 = **$3.94**

25. b. total aid ÷ number of students = aid per student
last year's aid − decrease = total aid
1,126,200 − 462,000 = 664,200 total aid
664,200 ÷ 820 students = **$810**

26. c. original price − savings = sale price
Since the item is over 40 days old, the discount is 75%.

$$\frac{savings}{original\ price} = \frac{discount}{100\%}$$

$$\frac{s}{\$18} = \frac{75\%}{100\%}$$

$100\ s = 18 \times 75$
$s = 1350 \div 100 = \$13.50$
$18 - 13.50 =$ sale price
$4.50 = sale price

27. e. original price − savings = sale price
red suit: $40 − (.25 × $40) = $30
floral print suit: $30 − (.10 × $30) = $27
violet suit: $45 − (.40 × $45) = $27
striped suit: $28 − (0 × $28) = $28
black suit: $60 − (.75 × $60) = **$15**

28. e. You don't know the price of the sweater.

29. c. total bill = customer charge + kWh (delivery charge + supplier charge)
total bill = **$5.81 + 445 kWh ($0.047 + $0.032)**

30. d. cost of digital camera + cost of Zip disk = cost of SLR camera + rolls of film (cost of roll + cost of developing)
$499 + $10 = $249 + r ($4 + $16)
$260 = 20r$
$r = 13$ rolls of film
pictures = rolls of film × pictures per roll = 13 × 36 = **468 pictures**

Pages 12–13

1. How much snow fell during the entire winter?

2. What is the total cooking time?

3. Find the cost of parking at the meter.

4. How many years did Joe serve in prison?

5. Find the number of flats she can buy.

6. How much did the city of Eugene receive?

Answers may vary.

7. How many coats can be made from a roll of fabric? What percent of the roll is used to make each coat?

8. How many grams of acetaminophen are in a normal dose? What is the maximum grams of acetaminophen that a normal adult can take in a day? How many normal doses of Tylenol can an adult take in a 24-hour period?

9. What was the price reduction? By what percent was the price reduced?

10. How many yards of fabric can be produced in an hour? How many yards of fabric can be produced in a week? If the factory closes on Sunday, how many yards of fabric can be produced in a week?

11. How many more cars were sold during Presidents' Week? By what percent did sales increase during Presidents' Week?

12. How much would you have to pay for a super pretzel and a soda? (You could use any combination of items from the menu.) If you bought a hot dog, chips, and a soda and paid for it with a $10 bill, how much change would you get? If you bought 2 hamburgers, 1 hot dog, 2 pizza slices, and 5 sodas for your family and paid for the food with a $20 bill, how much change would you get?

Page 15

1. numbers and labels: 124 commuters, 119 commuters
label of answer: commuters

2. numbers and labels: 14 potatoes, 5 lb
label of answer: pounds (or lb)

3. numbers and labels: $0.06, $1.47
label of answer: $

4. numbers and labels: 2 qt, 1 qt
label of answer: quarts or qt

5. numbers and labels: $38, $329
label of answer: $

6. numbers and labels: 60 pages, 250 pages
label of answer: pages

Page 17

1. *given information:* 22 years, 20 years, 23 years
   necessary information: 22 years, 20 years
   You are only comparing Mona's age with her sister's age. Therefore, the boyfriend's age is not needed.

2. *given information:* $186, $167, 2 children
   necessary information: $186, $167
   You do not need to know Rena's number of children in order to find the total amount of assistance she receives.

3. *given information:* 3 times, 20-year-old, 10 hours
   necessary information: 3 times, 10 hours
   You do not need to use Laura's age to find out how many hours Marilyn works.

4. *given information:* 7-year-old, $43, $39, $40, $31, first 2 months
   necessary information: $43, $39
   The chart shows information for 4 months; you are asked for information about the first months—January and February. The age of the car is not necessary.

5. *given information:* $2,460, $35,800, $1.20
   necessary information: $2,460, $1.20
   The money made on shoes is not needed to determine how many gallons of oil were bought.

6. *given information:* 45, 8 people, $\frac{1}{2}$ of the family
   necessary information: 8 people, $\frac{1}{2}$ of the family
   Erma's age is not needed to answer the question, "For how many people does Jack cook?"

7. *given information:* 4,700 workers, 3,900, 700 of the employees
   necessary information: 4,700, 700
   In finding the number of employees currently working, the total number of skilled workers is not necessary information.

8. *given information:* $1.49, 12-ounce, 64 oz
   necessary information: 12 ounce, 64 oz
   The cost of the cola is not needed to find the number of glasses Maritza can fill.

Page 19

1. plus
2. and, altogether
3. added, extra
4. increase
5. more, altogether
6. in all

Page 21

1. altogether
   5 inches
   + 23 inches
   28 inches

2. increased
   $1.80
   + 0.20
   $2.00

3. total
   3,500 pounds
   + 720 pounds
   4,220 pounds

4. larger than
   2 rooms
   + 3 rooms
   5 rooms

5. and
   $121,460
   + $ 89,742
   $211,202

6. both, and
   $529
   + $449
   $978

Page 22

1. cheaper than
2. decrease
3. less than
4. difference
5. reduced

Pages 23–24

1. left
   105 homes
   − 36 homes
   69 homes

2. change
   $20
   − $16
   $ 4

3. left
   $361
   − $325
   $ 36

4. difference
   $12,635
   − $ 7,849
   $ 4,786

5. fallen
   86 degrees
   − 12 degrees
   74 degrees

6. more than
   161 pounds
   − 104 pounds
   57 pounds

7. closer
   420 miles
   − 140 miles
   280 miles

8. decrease
   $121
   − $ 46
   $ 75

9. remained, lost
   460 books
   − 133 books
   327 books

10. less
    21 pounds
    − 17 pounds
    4 pounds

Pages 26–27

1. and, altogether
   <u>add</u>
      86 books
   + 53 books
   139 books

2. more than
   <u>subtract</u>
      31 rings
   – 15 rings
   16 rings

3. total
   <u>add</u>
     564 votes
   + 365 votes
   929 votes

4. decrease
   <u>subtract</u>
     421 units
   – 253 units
   168 units

5. farther
   <u>subtract</u>
    10,000 miles
   – 3,000 miles
   7,000 miles

6. increased
   <u>add</u>
     $530
   +   35
   $565

7. more than
   <u>subtract</u>
     $42
   – 26
   $16

Pages 29–30

1. rose
2. rose
3. increase
4. decrease
5. less
6. less
7. lowered
8. raised
9. more

Pages 32–33

1. b
2. a
3. b
4. b
5. a
6. a
7. b
8. b

Pages 35–36

1. subtract: 31 students
             – 3 students
             28 students

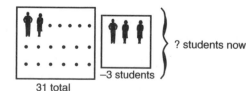

2. add: $483 1998 taxes
         + $39 difference
         $522 1999 taxes

3. subtract: 12,000 gallons
            – 3,500 gallons
            8,500 gallons

4. add: $120 inkjet printer
         + $359 difference
         $479 laser printer

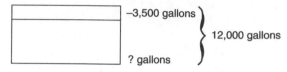

5. subtract: $359 laser printer
            – $120 difference
            $239 inkjet printer

6. add: 61 degrees at 11 P.M.
         + 13-degree difference
         74 degrees at 6 P.M.

7. subtract: 61 degrees at 6 P.M.
           – 13-degree decrease
           48 degrees at 11 P.M.

8. subtract: 1,412 graduates
       −   457 graduates living
       955 graduates who died

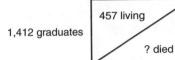

9. subtract: $6,000 loan
       − $3,800 paid back
       $2,200 owed

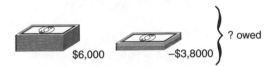

10. add: 2,600-pound truck
       + 1,000-pound load
       3,600 pounds total

11. add and compare: 24 pounds
       + 42 pounds
       66 pounds

66 pounds is less than 70 pounds. The total weight did not exceed the weight limit.

12. add: 790,000 legal copies
       + 600,000 illegal copies
       1,390,000 copies

Pages 38–39

1. *necessary information:* 20-cent stamp, quarter
amount paid − price of stamp = amount of change
25 cents − 20 cents = amount of change
5 cents = amount of change
    25 cents
− 20 cents
    5 cents

2. *necessary information:* 32 miles, 51 miles
total miles driven − commuting distance = additional driving
51 miles − 32 miles = additional driving
19 miles = additional driving
    51 miles
− 32 miles
    19 miles

3. *necessary information:* 134 tickets, 172 Saturday tickets
ticket sales needed to break even − tickets sold = tickets to be sold
172 tickets − 134 tickets = tickets to be sold
38 tickets = tickets to be sold
    172 tickets
− 134 tickets
    38 tickets

4. *necessary information:* $3,300, $1,460
price of car − savings = loan
$3,300 − $1,460 = loan
$1,840 = loan
    $3,300
−   1,460
    $1,840

5. *necessary information:* 150 names, 119 names
names needed − names on first day = more names
150 names − 119 names = more names
31 names = more names
    150 names
− 119 names
    31 names

6. *necessary information:* 47 pounds, 119 pounds
new weight + weight loss = original weight
119 pounds + 47 pounds = original weight
166 pounds = original weight
    119 pounds
+   47 pounds
    166 pounds

7. *necessary information:* $13, $168
new food stamp allotment + reduction = original amount
$168 + $13 = original amount
$181 = original amount
    $168
+   13
    $181

8. *necessary information:* $465, $379
original price − sales price = savings
$465 − $379 = savings
$86 = savings
    $465
−   379
    $86

9. *necessary information:* $213, $185
dollars withheld − amount owed = refund
$213 − $185 = refund
$28 = refund
    $213
−   185
    $28

10. *necessary information:* 3,500 cars, 8,200 cars
    new production + production cut = original
    production
    8,200 cars + 3,500 cars = original production
    11,700 cars = original production

    8,200 cars
    + 3,500 cars
    11,700 cars

11. *necessary information:* $28,682, $27,991
    dollars earned – dollars spent = dollars saved
    $28,682 – $27,991 = dollars saved
    $691 = dollars saved

    $28,682
    – $27,991
    $691

12. *necessary information:* 1,423 gallons, 1,289 gallons
    gallons Bertha + gallons Calico = total gallons
    1,423 gallons + 1,289 gallons = total gallons
    2,712 gallons = total gallons

    1,423 gallons
    + 1,289 gallons
    2,712 gallons

13. *necessary information:* 72,070 seats, 58,682 people
    total seats – people attended = empty seats
    72,070 seats – 58,682 people = empty seats
    13,388 seats = empty seats

    72,070 seats
    – 58,682 people
    13,388 seats

14. *necessary information:* 49 days, 56 days
    days spinach + days green beans = total days
    49 days + 56 days = total days
    105 days = total days

    49 days
    + 56 days
    105 days

Page 43

1. c   4. d
2. b   5. f
3. a   6. e

Page 45

1. f   4. d
2. e   5. a
3. c   6. b

Pages 46–47

Estimations may vary.

1. **b.**
   7.1 miles – 6.3 miles = **.8 mile**
   estimation: 7 miles – 6 miles = **1 mile**

2. **a.**
   9.4 gallons + 14.7 gallons = **24.1 gallons**
   estimation: 9 gallons + 15 gallons = **24 gallons**

3. **a.**
   1.42 pounds + .98 pound = **2.40 pounds**
   estimation: 1 pound + 1 pound = **2 pounds**

4. **b.**
   200.15 mph – 198.7 mph = **1.45 mph**
   estimation: 200 mph – 199 mph = **1 mph**

5. **b.**
   9.1% – 7.9% = **1.2%**
   estimation: 9% – 8% = **1%**

Pages 48–49

1. $2.60
   + $0.25
   $2.85

2. 0.70 gram
   – 0.55 gram
   0.15 gram

3. .342
   – .083
   .259   drop

4. $313.50
   – $126.13
   $187.37

5. 1.60 inches
   – 0.05 inch
   1.55 inches

6. $46.65
   + 23.35
   $70.00

7. 0.080 inch
   + 0.015 inch
   0.095 inch

8. 126.40 tons
   – 18.64 tons
   107.76 tons

9. 1.80 milliliters
   − 1.45 milliliters
   0.35 milliliter

┌─────────┐
│   ?   │ ⎫
├─────────┤ ⎬ 1.8 ml
│ 1.45 ml │ ⎭
└─────────┘

10. 2.64 pounds
   − 2.10 pounds
   0.54 pound

┌─────────┐
│   ?   │ ⎫
├─────────┤ ⎬ 2.64 lb
│ 2.1 lb │ ⎭
└─────────┘

Pages 50–51

1. *necessary information:* 0.6 ounce, 2.4 ounces
new weight + reduction = original weight
2.4 ounces + 0.6 ounce = original weight
3.0 ounces = original weight
   2.4 ounces
 + 0.6 ounce
   3.0 ounces

2. *necessary information:* $3.38, $10.00
amount paid − lunch cost = change
$10.00 − $3.38 = change
$6.62 = change
   $10.00
 − $3.38
   $6.62

3. *necessary information:* 3.94 pounds, 4.68 pounds
first chicken + second chicken = total weight
3.94 pounds + 4.68 pounds = total weight
8.62 pounds = total weight
   3.94 pounds
 + 4.68 pounds
   8.62 pounds

4. *necessary information:* 23,172.3 miles, 23,391.4 miles
reading at end − reading at start = length of trip
23,391.4 miles − 23,172.3 miles = length of trip
219.1 miles = length of trip
   23,391.4 miles
 − 23,172.3 miles
   219.1 miles

5. *necessary information:* $0.40, $0.65
bus + subway = 1-way trip
$0.40 + $0.65 = 1-way trip
$1.05 = 1-way trip
   $0.40
 + $0.65
   $1.05

6. *necessary information:* $341.98, $335.26
Massachusetts cost − New Hampshire cost = savings
$341.98 − $335.26 = savings
$6.72 = savings
   $341.98
 − $335.26
   $6.72

7. *necessary information:* 14.36 seconds, 13.9 seconds
first 100 meters + second 100 meters = total time
14.36 seconds + 13.9 seconds = total time
28.26 seconds = total time
   14.36 seconds
 + 13.90 seconds
   28.26 seconds

8. *necessary information:* 966 bottles, 50 bottles
bottles by 1st line − number of bottles fewer = bottles by 2nd line
966 bottles − 50 bottles = bottles by 2nd line
916 bottles = bottles by 2nd line
   966 bottles
 − 50 bottles
   916 bottles

Pages 52–53

1. b
$$28\tfrac{1}{2} = \;\;28\tfrac{2}{4} \text{ inches}$$
$$+ 31\tfrac{1}{4} = + 31\tfrac{1}{4} \text{ inches}$$
$$\mathbf{59\tfrac{3}{4} \text{ inches}}$$

2. a
$$1\tfrac{2}{3} \text{ cups}$$
$$+ 1\tfrac{1}{3} \text{ cups}$$
$$2\tfrac{3}{3} = \mathbf{3 \text{ cups}}$$

3. a
$$71\tfrac{1}{4} = \;\;71\tfrac{1}{4} = \;\;70\tfrac{5}{4} \text{ pounds}$$
$$- 62\tfrac{1}{2} = - 62\tfrac{2}{4} = - 62\tfrac{2}{4} \text{ pounds}$$
$$\mathbf{8\tfrac{3}{4} \text{ pounds}}$$

4. b
$$23\tfrac{1}{4} = \;\;22\tfrac{5}{4} \text{ inches}$$
$$- 18\tfrac{3}{4} = - 18\tfrac{3}{4} \text{ inches}$$
$$4\tfrac{2}{4} \text{ inches} = \mathbf{4\tfrac{1}{2} \text{ inches}}$$

Page 54

1. $4 = \;\;3\tfrac{8}{8}$ inches
   $-\tfrac{5}{8} = -\tfrac{5}{8}$ inch
   $\mathbf{3\tfrac{3}{8} \text{ inches}}$

2. $8\tfrac{1}{2} = \;\;8\tfrac{2}{4} = \;\;7\tfrac{6}{4}$ hours
   $-1\tfrac{3}{4} = -1\tfrac{3}{4} = -1\tfrac{3}{4}$ hours
   $\mathbf{6\tfrac{3}{4} \text{ hours}}$

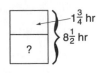

3. $6\tfrac{1}{2} = \;\;6\tfrac{2}{4}$ hours
   $+ \tfrac{3}{4} = + \tfrac{3}{4}$ hour
   $6\tfrac{5}{4} = \mathbf{7\tfrac{1}{4} \text{ hours}}$

4. $2 = 1\frac{2}{2}$ inches

$\quad \underline{-\frac{1}{2} = -\ \frac{1}{2}}$ inch

$\qquad\qquad 1\frac{1}{2}$ **inches**

$\frac{1}{2}$ in.

2 in.

5. $2\frac{1}{2} = 2\frac{3}{6} = 1\frac{9}{6}$ cups

$\quad \underline{-1\frac{2}{3} = -1\frac{4}{6} = -1\frac{4}{6}}$ cups

$\qquad\qquad\qquad\qquad \frac{5}{6}$ **cup**

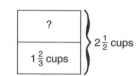

?

$1\frac{2}{3}$ cups $\Big\}$ $2\frac{1}{2}$ cups

6. $\frac{3}{4} = \frac{12}{16}$ inch

$\quad \underline{-\frac{1}{16} = -\ \frac{1}{16}}$ inch

$\qquad\qquad \frac{11}{16}$ **inch**

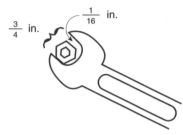

$\frac{3}{4}$ in.

$\frac{1}{16}$ in.

Page 55

1. bowl – rum = other ingredients

3 quarts $- 1\frac{1}{4}$ quarts = other ingredients

$1\frac{3}{4}$ quarts = other ingredients

$\quad\ 3\ = \ 2\frac{4}{4}$ quarts

$\underline{-1\frac{1}{4} = -1\frac{1}{4}}$ quarts

$\qquad\quad 1\frac{3}{4}$ **quarts**

2. original length – new length = amount taken off

$34\frac{1}{2}$ inches $- 32\frac{3}{4}$ inches = amount taken off

$1\frac{3}{4}$ inches = amount taken off

$\quad 34\frac{1}{2} = \ 33\frac{6}{4}$ inches

$\underline{-32\frac{3}{4} = -32\frac{3}{4}}$ inches

$\qquad\qquad 1\frac{3}{4}$ **inches**

3. original amount – amount sold = amount left

$6\frac{1}{2}$ yards $- 3\frac{2}{3}$ yards = amount left

$2\frac{5}{6}$ yards = amount left

$\quad 6\frac{1}{2} = \ 5\frac{9}{6}$ yards

$\underline{-3\frac{2}{3} = -3\frac{4}{6}}$ yards

$\qquad\qquad 2\frac{5}{6}$ **yards**

4. first week + second week = total wood

$\frac{1}{8}$ cord $+ \frac{1}{12}$ cord = total wood

$\frac{5}{24}$ cord = total wood

$\quad\ \frac{1}{8} = \ \frac{3}{24}$ cord

$\underline{+\frac{1}{12} = \ \frac{2}{24}}$ cord

$\qquad\quad \frac{5}{24}$ **cord**

5. total cloth – hem = length of drapes

$62\frac{1}{2}$ inches $- \frac{3}{4}$ = length of drapes

$61\frac{3}{4}$ inches = length of drapes

$\quad 62\frac{1}{2} = \ 61\frac{6}{4}$ inches

$\underline{-\ \frac{3}{4} = -\ \frac{3}{4}}$ inch

$\qquad\quad 61\frac{3}{4}$ **inches**

Page 57

Letters in equations and setup of equations may vary.

1. hits in February – hits in January = difference

4,348 hits – 2,917 hits = h

1,431 hits = h

2. payment – cost = change

$20.00 – $9.98 = c

$10.02 = c

3. calories of cheese pizza + calories of pepperoni = calories of pepperoni pizza

$\quad\ 134$ calories $+ c = \ 149$ calories

$\underline{-134}$ calories $\quad = -134$ calories

$\qquad\qquad\qquad\quad c = \ $ **15 calories**

4. length of board – length needed = length left over

$3\frac{1}{2}$ feet $- 1\frac{3}{4}$ feet $= f$

$1\frac{3}{4}$ **feet** $= f$

5. mg acetaminophen + mg aspirin + mg caffeine = mg active ingredients

250 mg + 250 mg + 65 mg = m

565 mg = m

6. current pile height – standard pile height = amount sheared off

$\frac{5}{8}$ in. $- \frac{7}{16}$ in. $= s$

$\frac{3}{16}$ **in.** $= s$

Pages 58–60

1. b. *subtract*: $\qquad$ 55,572 nails

$\qquad\qquad\qquad \underline{-\ \ 1,263}$ nails

$\qquad\qquad\qquad$ **54,309 nails**

2. b. *subtract*: $\qquad$ $75.62

$\qquad\qquad\qquad \underline{-\$38.56}$

$\qquad\qquad\qquad$ **$37.06**

3. d. *subtract*: $\qquad$ 19.1 miles per gallon

$\qquad\qquad\qquad \underline{-\ 16.2}$ miles per gallon

$\qquad\qquad\qquad$ **2.9 miles per gallon**

4. d. *add*: $\qquad$ $1,800

$\qquad\qquad\quad \underline{+\ \$6,400}$

$\qquad\qquad\quad$ **$8,200**

5. c. *subtract:*

$$\frac{1}{4} \text{ pound} = \frac{4}{16} \text{ pound}$$
$$-\frac{3}{16} \text{ pound} = -\frac{3}{16} \text{ pound}$$
$$\frac{1}{16} \textbf{ pound}$$

6. c. *subtract:*

$$0.60 \text{ gram}$$
$$-0.47 \text{ gram}$$
$$\textbf{0.13 gram}$$

7. a. *add:*

$$\$391,445$$
$$+\$528,555$$
$$\textbf{\$920,000}$$

8. b. *add:*

$$26\frac{3}{4} \text{ inches} = 26\frac{6}{8} \text{ inches}$$
$$+2\frac{3}{8} \text{ inches} = +2\frac{3}{8} \text{ inches}$$
$$28\frac{9}{8} \text{ inches} = \textbf{29}\frac{1}{8} \textbf{ inches}$$

9. e. *add:*

$$\frac{5}{8} \text{ cord} = \frac{15}{24} \text{ cord}$$
$$+\frac{1}{12} \text{ cord} = +\frac{2}{24} \text{ cord}$$
$$\frac{17}{24} \textbf{ cord}$$

10. e. *subtract:*

$$5.15 \text{ tubes}$$
$$-3.40 \text{ tubes}$$
$$\textbf{1.75 tubes}$$

11. b. *add:*

$$\frac{1}{2} \text{ radioactivity} = \frac{8}{16} \text{ radioactivity}$$
$$+\frac{7}{16} \text{ radioactivity} = +\frac{7}{16} \text{ radioactivity}$$
$$\frac{15}{16} \textbf{ radioactivity}$$

12. d. *add:*

$$2.77 \text{ grams}$$
$$+0.03 \text{ gram}$$
$$\textbf{2.80 grams}$$

Pages 61–62

1. times
2. per
3. total
4. twice
5. by
6. area
7. multiplied, times
8. per
9. per

Page 63

1. *necessary information:* 92 cents, twice
key word: twice

$$92 \text{ cents}$$
$$\times 2$$
$$184 \text{ cents or } \textbf{\$1.84}$$

2. *necessary information:* 4 sets, 6 strings per set
key word: per

$$4 \text{ sets}$$
$$\times 6 \text{ strings per set}$$
$$\textbf{24 strings}$$

3. *necessary information:* 5 times a day, week
key word: times

$$5 \text{ times a day}$$
$$\times 7 \text{ days (week)}$$
$$\textbf{35 times}$$

4. *necessary information:* $75 per hour, 3 hours
key word: per

$$\$75 \text{ per hour}$$
$$\times 3 \text{ hours}$$
$$\textbf{\$225}$$

Page 64

1. share, equally
2. per
3. average, each
4. shared, equally, each

Page 65

1. *necessary information:* 12 oz, 4 children
key words: shared equally, each
12 oz ÷ 4 children

$$4\overline{)12} \quad \textbf{3 oz}$$

2. *necessary information:* 60-minute hockey game, 3 equal periods
key words: divided, equal (periods)
60-minute hockey game ÷ 3 equal periods

$$3\overline{)60} \quad \textbf{20 minutes}$$

3. *necessary information:* $156, 12 monthly payments
key word: each
$156 ÷ 12 monthly payments

$$12\overline{)156} \quad \textbf{\$13}$$

4. *necessary information:* 24 mints, 96 cents
key word: each
96 cents ÷ 24 mints

$$24\overline{)96} \quad \textbf{4 cents}$$

5. *necessary information:* $3, $4,629
key words: each, even
$4,629 ÷ $3

$$3\overline{)4,629} \quad \textbf{1,543 tickets}$$

Page 67

1. 5,500	72,600	4,500
2. 43,000	9,900	7,000
3. 380	449.3	.772
4. 48	29.72	1.092

Page 68

1. *division key words:* divided equally, each

$$4\overline{)8} \quad \begin{array}{c}\textbf{2 pieces}\end{array}$$

2. *division key words:* each, divided equally

$$5\overline{)90,000} \quad \begin{array}{c}\textbf{\$18,000}\end{array}$$

3. *multiplication key words:* per, total

$$\begin{array}{r} 22 \\ \times\ 15 \\ \hline \textbf{\$330} \end{array}$$

4. *multiplication key word:* twice

$$\begin{array}{r} 47 \\ \times\ 2 \\ \hline \textbf{\$94} \end{array}$$

5. *division key words:* each, average

$$5\overline{)225} \quad \begin{array}{c}\textbf{45 points}\end{array}$$

6. *multiplication key word:* per

$$\begin{array}{r} 28 \\ \times\ 15 \\ \hline \textbf{420 miles} \end{array}$$

Page 71

1. part: 78 cartons
   part: 50 cups
$$\begin{array}{r} 78 \\ \times\ 50 \\ \hline 3,900 \end{array}$$
   total: **3,900 cups**

   ? total
   78 part   50 part

2. part: 30 days
   part: $16 a day
$$\begin{array}{r} 30 \\ \times\ 16 \\ \hline \$480 \end{array}$$
   total: **$480**

   ? total
   30 part   16 part

3. part: 780,000 copies
   part: 140 pages
$$\begin{array}{r} 780,000 \\ \times\ 140 \\ \hline 109,200,000 \end{array}$$
   total: **109,200,000 pages**

   ? total
   140 part   780,000 part

4. part: 21 gallons
   part: 18 miles per gallon
$$\begin{array}{r} 21 \\ \times\ 18 \\ \hline 378 \end{array}$$
   total: **378 miles**

   ? total
   21 part   18 part

5. part: 365 days in a year
   total: 46,720 people
$$365\overline{)46,720} \quad \begin{array}{c}128\end{array}$$
   part: **128 people a day**

   46,720 total
   365 part   ? part

6. part: 150 nails per box
   total: 68,400 nails
$$150\overline{)68,400} \quad \begin{array}{c}456\end{array}$$
   part: **456 boxes**

   68,400 total
   150 part   ? part

Pages 73–76

1. a.
$$\begin{array}{r} 380 \text{ lab animals} \\ \times\ 5 \text{ oz per animal} \\ \hline 1,900 \text{ oz} \end{array}$$

2. b.
$$\$6 \text{ per person}\overline{)\$3,102} \quad \begin{array}{c}\textbf{517 people}\end{array}$$

3. b.
$$13 \text{ gallons}\overline{)260 \text{ miles}} \quad \begin{array}{c}\textbf{20 miles per gallon}\end{array}$$

4. a.
$$\begin{array}{r} 48 \text{ plants per row} \\ \times\ 8 \text{ rows} \\ \hline \textbf{384 plants} \end{array}$$

5. a.
$$400 \text{ blouses}\overline{)\$6,000} \quad \begin{array}{c}\textbf{\$15 per blouse}\end{array}$$

6. divided equally among

7. each

8. per square mile

9. On the average, how many hamburgers were sold per hour?

10. people

Pages 78–79

1. miles × minutes per mile = total minutes
   6.2 miles × 6.5 minutes per mile = total minutes
$$\begin{array}{r} 6.2 \\ \times\ 6.5 \\ \hline 40.30 \end{array}$$
   total: **40.3 minutes**

   ? total
   6.2 part   6.5 part

2. total price ÷ number of pounds = price per pound
$23.20 ÷ 40 pounds = price per pound

$$\frac{.58}{40)23.20}$$

part: **$0.58**

23.20
total

40	?
part	part

3. total miles ÷ gallons = miles per gallon
159.75 miles ÷ 7.1 gallons = miles per gallon

$$\frac{22.5}{7.1)159.7.5}$$

part: **22.5 miles per gallon**

159.75
total

7.1	?
part	part

4. number of pounds × price per pound = total price
4.67 pounds × $2.29 per pound = total price

$$\begin{array}{r} 4.67 \\ \times\ 2.29 \\ \hline 10.6943 \end{array}$$

total: **$10.69**

?
total

4.67	2.29
part	part

5. food bill ÷ roommates = price per roommate
$372.36 ÷ 4 roommates = price per roommate

$$\frac{93.09}{4)372.36}$$

part: **$93.09 per roommate**

372.36
total

4	?
part	part

6. total commission ÷ hours = hourly commission
$118.56 ÷ 8 hours = hourly commission

$$\frac{14.82}{8)118.56}$$

part: **$14.82 per hour**

118.56
total

8	?
part	part

7. total $ collected ÷ cost per ride = number of passengers
$167.45 ÷ 0.85 = number of passengers

$$\frac{197.}{.85)167.45}$$

part: **197 passengers**

167.45
total

0.85	?
part	part

8. cost per day × number of days = total cost
$3.65 × 239 = total cost

$$\begin{array}{r} \$3.65 \\ \times\ 239 \\ \hline \$872.35 \end{array}$$

total: **$872.35**

?
total

3.65	239
part	part

Page 82

1. *necessary information:* $\frac{2}{3}$ of, 36 inches

fraction (of) × total precipitation = inches of rain

$$\frac{2}{3} \times 36 \text{ inches} = \frac{2}{1\cancel{3}} \times \frac{\cancel{36}^{12}}{1} = \textbf{24 inches}$$

2. *necessary information:* $\frac{7}{8}$ of, 23,352 accidents

fraction (of) × total accidents = accidents in urban areas

$$\frac{7}{8} \times 23{,}352 \text{ accidents} =$$

$$\frac{7}{1\cancel{8}} \times \frac{\cancel{23{,}352}^{2{,}919}}{1} = \textbf{20,433 accidents}$$

3. *necessary information:* $7\frac{2}{3}$ pounds, 10 boxes

pounds per box × number of boxes = total pounds

$$7\frac{2}{3} \times 10 = \frac{23}{3} \times \frac{10}{1} = \frac{230}{3} = \textbf{76}\frac{2}{3}\textbf{ pounds}$$

4. *necessary information:* $\frac{2}{3}$ can of, 12 days

cans per day × number of days = total cans

$$\frac{2}{3} \text{ can per day} \times 12 \text{ days} = \frac{2}{1\cancel{3}} \times \frac{\cancel{12}^4}{1} = \textbf{8 cans}$$

5. *necessary information:* $1\frac{1}{8}$ ounces, $3\frac{1}{2}$ candy bars

ounces per candy bar × number of candy bars = total ounces

$$1\frac{1}{8} \text{ ounces per candy bar} \times 3\frac{1}{2} \text{ candy bars} =$$

$$\frac{9}{8} \times \frac{7}{2} = \frac{63}{16} = \textbf{3}\frac{15}{16}\textbf{ ounces}$$

6. *necessary information:* $\frac{1}{2}$ cup, $\frac{1}{2}$ of a load

fraction (of) × cups per load = detergent needed

$$\frac{1}{2} \times \frac{1}{2} \text{ cup} = \frac{1}{2} \times \frac{1}{2} = \frac{1}{4}\textbf{ cup}$$

7. *necessary information:* $\frac{2}{5}$ of, $\frac{1}{4}$ pound

fraction (of) × weight of hamburger = amount of fat

$$\frac{2}{5} \times \frac{1}{4} \text{ pound} = \frac{\cancel{2}^1}{5} \times \frac{1}{\cancel{4}_2} = \frac{1}{10}\textbf{ pound}$$

8. *necessary information:* 17,000 miles per hour, $2\frac{1}{2}$ hours

miles per hour × hours = miles

$$17{,}000 \times 2\frac{1}{2} =$$

$$\frac{\cancel{17{,}000}^{8{,}500}}{1} \times \frac{5}{\cancel{2}_1} = \textbf{42,500 miles}$$

Page 84

1. *necessary information:* $22\frac{1}{2}$ inches, $\frac{5}{8}$ inch
   depth of box ÷ thickness of each book = number of books
   $22\frac{1}{2}$ inches ÷ $\frac{5}{8}$ inch per book = $\frac{\overset{9}{\cancel{45}}}{\underset{1}{\cancel{2}}} \times \frac{\overset{4}{\cancel{8}}}{\cancel{5}_{1}}$ = **36 books**

2. *necessary information:* 13 people, $6\frac{1}{2}$ pounds
   total amount of meat ÷ number of people = size of a portion
   $6\frac{1}{2}$ pounds ÷ 13 people = $\frac{\overset{1}{\cancel{13}}}{2} \times \frac{1}{\cancel{13}_{1}} = \frac{1}{2}$ **pound per person**

3. *necessary information:* $2\frac{1}{4}$ feet, $265\frac{1}{2}$ feet
   total ribbon ÷ ribbon per book = number of books
   $265\frac{1}{2}$ feet ÷ $2\frac{1}{4}$ feet = $\frac{531}{2} \div \frac{9}{4}$ =
   $\frac{\overset{59}{\cancel{531}}}{\underset{1}{\cancel{2}}} \times \frac{\overset{2}{\cancel{4}}}{\cancel{9}_{1}} = \frac{59}{1} \times \frac{2}{1}$ = **118 books**

4. *necessary information:* $8\frac{1}{2}$ pounds, $\frac{1}{2}$ pound
   total mashed potatoes ÷ potatoes per serving = number of servings
   $8\frac{1}{2}$ pounds ÷ $\frac{1}{2}$ pound per serving = $\frac{17}{\cancel{2}_{1}} \times \frac{\cancel{2}^{1}}{1}$ = **17 servings**

5. *necessary information:* $9\frac{3}{4}$ ounces, 3 equal servings
   total ounces of peaches ÷ number of servings = size of each serving
   $9\frac{3}{4}$ ounces ÷ 3 servings = $\frac{\overset{13}{\cancel{39}}}{4} \times \frac{1}{\cancel{3}_{1}}$ =
   $\frac{13}{4} = 3\frac{1}{4}$ **ounces per serving**

6. *necessary information:* 12 feet, $1\frac{1}{2}$-foot wide sections
   total width ÷ width of section = number of sections
   12 feet ÷ $1\frac{1}{2}$ feet per section = $\frac{\overset{4}{\cancel{12}}}{1} \times \frac{2}{\cancel{3}_{1}}$ = **8 sections**

Pages 85–86

1. *necessary information:* 12 hours, $\frac{3}{4}$ hour
   total hours ÷ length of a session = number of sessions
   12 hours ÷ $\frac{3}{4}$ hour per session = $\frac{\overset{4}{\cancel{12}}}{1} \times \frac{4}{\cancel{3}_{1}}$ = **16 sessions**

2. *necessary information:* $\frac{2}{3}$ hour, 30 cars
   time per car × number of cars = total hours
   $\frac{2}{3}$ hour per car × 30 cars = $\frac{2}{\cancel{3}_{1}} \times \frac{\overset{10}{\cancel{30}}}{1}$ = **20 hours**

3. *necessary information:* $\frac{2}{3}$ hour, 24 hours
   total hours ÷ time per car = number of cars
   24 hours ÷ $\frac{2}{3}$ hour per car = $\frac{\overset{12}{\cancel{24}}}{1} \times \frac{3}{\cancel{2}_{1}}$ = **36 cars**

4. *necessary information:* $\frac{2}{3}$ of, 26,148 microwave ovens
   fraction (of) × total ovens = defective ovens
   $\frac{2}{3} \times 26,148$ microwave ovens = $\frac{2}{\cancel{3}_{1}} \times \frac{\overset{8716}{\cancel{26,148}}}{1}$ =
   17,432 microwave ovens

5. *necessary information:* 6 hikers, $4\frac{1}{2}$ pounds
   total chocolate ÷ number of hikers = chocolate per hiker
   $4\frac{1}{2}$ pounds ÷ 6 hikers = $\frac{9}{2} \div \frac{6}{1} = \frac{\overset{3}{\cancel{9}}}{2} \times \frac{1}{\cancel{6}_{2}} = \frac{3}{4}$ **pound per hiker**

6. *necessary information:* $1\frac{1}{2}$ teaspoons, double
   original number of teaspoons × 2 = doubled recipe
   $1\frac{1}{2} \times 2 = \frac{3}{2} \times \frac{2}{1} = \frac{6}{2}$ = **3 teaspoons**

7. On average, how much toner is used for each copy?

8. What will the new phone company charge for her talks with her mother?

9. How much butter should she use for her new brownie recipe?

10. How much does a single fig bar weigh?

Pages 87–88

1. b. total weight ÷ weight of mayonnaise in each jar = number of jars
   40 pounds ÷ $\frac{5}{8}$ pound per jar = $\frac{\overset{8}{\cancel{40}}}{1} \times \frac{8}{\cancel{5}_{1}}$ = **64 jars**

2. c. total cost ÷ square feet = cost per square foot
   $20.80 ÷ 32 square feet = **$0.65**
   $$32\overline{)20.80}^{\,.65}$$

3. b. fraction (of) × total raffle tickets = raffle sales needed
   $\frac{1}{6} \times 3,000$ raffle tickets =
   $\frac{1}{\cancel{6}_{1}} \times \frac{\overset{500}{\cancel{3,000}}}{1}$ = **500 raffle tickets**

4. b. fraction (of) × winner's share = trainer's share
   $\frac{3}{5} \times \$17,490 = \frac{3}{\cancel{5}_{1}} \times \frac{\overset{3498}{\cancel{17,490}}}{1}$ = **$10,494**

5. d. total material ÷ material per apron = number of aprons
   $7\frac{1}{3}$ yards ÷ $\frac{2}{3}$ yard/apron =
   $\frac{22}{3} \div \frac{2}{3} = \frac{\overset{11}{\cancel{22}}}{\cancel{3}_{1}} \div \frac{\cancel{3}^{1}}{\cancel{2}_{1}}$ = **11 aprons**

6. d. total distance ÷ hours = miles per hour
   263.1 miles ÷ 4.5 hours = **58.5 miles per hour**
   $$4.5\overline{)263.1.}^{\,58.46\,=\,58.5}$$

7. **c.** total CDs ÷ CDs per box = number of boxes
1,410 CDs ÷ 30 CDs per box = **47 boxes**

$$30\overline{)1410} \quad 47$$

8. **b.** number of kilometers × mile per kilometer = number of miles
15 kilometers × 0.62 mile per kilometer = **9.3 miles**

$$\begin{array}{r} 15 \\ \times\ 0.62 \\ \hline 9.30 \end{array}$$

Page 89

2. $\dfrac{2\ \text{teachers}}{30\ \text{students}}$

3. $\dfrac{120\ \text{dollars}}{8\ \text{hours}}$

4. $\dfrac{38\ \text{miles}}{2\ \text{gallons}}$

5. $\dfrac{3\ \text{buses}}{114\ \text{commuters}}$

Page 92

1. $\dfrac{160\ \text{miles}}{5\ \text{hours}} = \dfrac{n\ \text{miles}}{10\ \text{hours}}$
$5 \times n = 160 \times 10$
$5n = 1{,}600$
$n = \dfrac{1{,}600}{5}$
$n = \textbf{320 miles}$

2. $\dfrac{12\ \text{cars}}{32\ \text{people}} = \dfrac{3\ \text{cars}}{n\ \text{people}}$
$12 \times n = 32 \times 3$
$12n = 96$
$n = \dfrac{96}{12}$
$n = \textbf{8 people}$

3. $\dfrac{n\ \text{dollars}}{8\ \text{quarters}} = \dfrac{6\ \text{dollars}}{24\ \text{quarters}}$
$24 \times n = 8 \times 6$
$24n = 48$
$n = \dfrac{48}{24}$
$n = \textbf{2 dollars}$

4. $\dfrac{42\ \text{pounds}}{n\ \text{chickens}} = \dfrac{14\ \text{pounds}}{4\ \text{chickens}}$
$14 \times n = 42 \times 4$
$14n = 168$
$n = \dfrac{168}{14}$
$n = \textbf{12 chickens}$

5. $\dfrac{28{,}928\ \text{people}}{8\ \text{doctors}} = \dfrac{n\ \text{people}}{1\ \text{doctor}}$
$8 \times n = 1 \times 28{,}928$
$8n = 28{,}928$
$n = \dfrac{28{,}928}{8}$
$n = \textbf{3,616 people}$

6. $\dfrac{\$24.39}{1\ \text{shirt}} = \dfrac{\$n}{6\ \text{shirts}}$
$1 \times n = \$24.39 \times 6$
$n = \textbf{\$146.34}$

7. $\dfrac{\$47.85}{3\ \text{shirts}} = \dfrac{\$n}{10\ \text{shirts}}$
$3 \times n = 47.85 \times 10$
$3n = 478.50$
$n = \dfrac{478.50}{3}$
$n = \textbf{\$159.50}$

8. $\dfrac{3\ \text{minutes}}{\frac{1}{2}\ \text{mile}} = \dfrac{n\ \text{minutes}}{5\ \text{miles}}$
$\frac{1}{2} \times n = 3 \times 5$
$\frac{1}{2}n = 15$
$n = 15 \div \frac{1}{2} = 15 \times 2$
$n = \textbf{30 minutes}$

9. $\dfrac{575\ \text{passengers}}{n\ \text{days}} = \dfrac{1{,}725\ \text{passengers}}{21\ \text{days}}$
$1{,}725 \times n = 575 \times 21$
$1{,}725n = 12{,}075$
$n = \dfrac{12{,}075}{1{,}725}$
$n = \textbf{7 days}$

10. $\dfrac{7\ \text{blinks}}{\frac{1}{10}\ \text{minute}} = \dfrac{n\ \text{blinks}}{10\ \text{minutes}}$
$\frac{1}{10} \times n = 7 \times 10$
$\frac{1}{10}n = 70$
$n = 70 \times 10$
$n = \textbf{700 blinks}$

Pages 94–95

1. *necessary information:* 7,800 people, 140,400 people
labels for proportion: $\dfrac{\text{shipments}}{\text{people}}$

$\dfrac{1\ \text{shipment}}{7{,}800\ \text{people}} = \dfrac{n\ \text{shipments}}{140{,}400\ \text{people}}$
$7{,}800 \times n = 1 \times 140{,}400$
$7{,}800n = 140{,}400$
$n = \dfrac{140{,}400}{7{,}800}$
$n = \textbf{18 shipments}$

2. *necessary information:* \$340, 24 hours
labels for proportion: $\dfrac{\$}{\text{hours}}$

$\dfrac{\$340}{1\ \text{hour}} = \dfrac{\$n}{24\ \text{hours}}$
$1 \times n = 340 \times 24$
$n = \textbf{\$8,160}$

3. *necessary information:* 11 ounces, 28 cans

labels for proportion: $\dfrac{\text{ounces}}{\text{cans}}$

$$\dfrac{11 \text{ ounces}}{1 \text{ can}} = \dfrac{n \text{ ounces}}{28 \text{ cans}}$$

$$1 \times n = 11 \times 28$$

$$n = \textbf{308 ounces}$$

4. *necessary information:* 52 words per minute, 26 minutes

labels for proportion: $\dfrac{\text{words}}{\text{minutes}}$

$$\dfrac{52 \text{ words}}{1 \text{ minute}} = \dfrac{n \text{ words}}{26 \text{ minutes}}$$

$$1 \times n = 52 \times 26$$

$$n = \textbf{1,352 words}$$

5. *necessary information:* 3,960 Band-Aids, 180 school days

labels for proportion: $\dfrac{\text{Band-Aids}}{\text{school days}}$

$$\dfrac{3,960 \text{ Band-Aids}}{180 \text{ school days}} = \dfrac{n \text{ Band-Aids}}{1 \text{ school day}}$$

$$180 \times n = 1 \times 3,960$$

$$180n = 3,960$$

$$n = \dfrac{3,960}{180}$$

$$n = \textbf{22 Band-Aids}$$

6. *necessary information:* 5,460 aspirin, 260 aspirin

labels for proportion: $\dfrac{\text{aspirin}}{\text{bottles}}$

$$\dfrac{260 \text{ aspirin}}{1 \text{ bottle}} = \dfrac{5,460 \text{ aspirin}}{n \text{ bottles}}$$

$$260 \times n = 1 \times 5,460$$

$$260n = 5,460$$

$$n = \dfrac{5,460}{260}$$

$$n = \textbf{21 bottles}$$

7. *necessary information:* 126 tons, 3-ton loads

labels for proportion: $\dfrac{\text{tons}}{\text{loads}}$

$$\dfrac{3 \text{ tons}}{1 \text{ load}} = \dfrac{126 \text{ tons}}{n \text{ loads}}$$

$$3 \times n = 1 \times 126$$

$$3n = 126$$

$$n = \dfrac{126}{3}$$

$$n = \textbf{42 loads}$$

8. *necessary information:* 18 feet, 3 feet in a yard

labels for proportion: $\dfrac{\text{feet}}{\text{yards}}$

$$\dfrac{3 \text{ feet}}{1 \text{ yard}} = \dfrac{18 \text{ feet}}{n \text{ yards}}$$

$$3 \times n = 18 \times 1$$

$$3n = 18$$

$$n = \dfrac{18}{3}$$

$$n = \textbf{6 yards}$$

9. *necessary information:* 26 water chestnuts, 3 cans

labels for proportion: $\dfrac{\text{water chestnuts}}{\text{cans}}$

$$\dfrac{26 \text{ water chestnuts}}{1 \text{ can}} = \dfrac{n \text{ water chestnuts}}{3 \text{ cans}}$$

$$1 \times n = 3 \times 26$$

$$n = \textbf{78 water chestnuts}$$

Page 97

1. *necessary information:* 25.4 millimeters, 100-millimeter

labels for proportion: $\dfrac{\text{millimeters}}{\text{inches}}$

$$\dfrac{25.4 \text{ millimeters}}{1 \text{ inch}} = \dfrac{100 \text{ millimeters}}{n \text{ inches}}$$

$$25.4 \times n = 1 \times 100$$

$$25.4n = 100$$

$$n = \dfrac{100}{25.4}$$

$$n = \textbf{3.94 inches}$$

2. *necessary information:* 2.2 pounds, 36 kilograms

labels for proportion: $\dfrac{\text{pounds}}{\text{kilograms}}$

$$\dfrac{2.2 \text{ pounds}}{1 \text{ kilogram}} = \dfrac{n \text{ pounds}}{36 \text{ kilograms}}$$

$$1 \times n = 2.2 \times 36$$

$$n = \textbf{79.2 pounds}$$

3. *necessary information:* 35.5 hours, \$8.62

labels for proportion: $\dfrac{\text{hours}}{\$}$

$$\dfrac{35.5 \text{ hours}}{\$ n} = \dfrac{1 \text{ hour}}{\$8.62}$$

$$1 \times n = 35.5 \times 8.62$$

$$n = \textbf{\$306.01}$$

4. *necessary information:* 1.09 yards, 880 yards

labels for proportion: $\dfrac{\text{yards}}{\text{meters}}$

$$\dfrac{1.09 \text{ yards}}{1 \text{ meter}} = \dfrac{880 \text{ yards}}{n \text{ meters}}$$

$$1.09 \times n = 1 \times 880$$

$$1.09n = 880$$

$$n = \dfrac{880}{1.09}$$

$$n = \textbf{807.34 meters}$$

5. *necessary information:* \$20.00, \$1.15

labels for proportion: $\dfrac{\$}{\text{gallons}}$

$$\dfrac{\$20}{n \text{ gallons}} = \dfrac{\$1.15}{1 \text{ gallon}}$$

$$1.15 \times n = 1 \times 20$$

$$1.15n = 20$$

$$n = \dfrac{20}{1.15}$$

$$n = \textbf{17.39 gallons}$$

6. *necessary information:* 1.61 kilometers, 55 miles per hour

labels for proportion:

$$\frac{\text{kilometers}}{\text{miles}} = \frac{\text{kilometers per hour}}{\text{miles per hour}}$$

(This proportion will work because the "per hour" appears on both top and bottom.)

$$\frac{1.61 \text{ kilometers}}{1 \text{ mile}} = \frac{n \text{ kilometers per hour}}{55 \text{ miles per hour}}$$

$$1 \times n = 1.61 \times 55$$

$$n = \textbf{88.55 kilometers per hour}$$

7. *necessary information:* 35.4 grams, 9,486 bars

labels for proportion: $\frac{\text{grams}}{\text{bars}}$

$$\frac{35.4 \text{ grams}}{1 \text{ Tiger Milk bar}} = \frac{n \text{ grams}}{9,486 \text{ Tiger Milk bars}}$$

$$1 \times n = 35.4 \times 9,486$$

$$n = \textbf{335,804.4 grams}$$

Page 99

1. *necessary information:* $\frac{1}{16}$ inch, 2 inches

labels for proportion: $\frac{\text{slices}}{\text{inches}}$

$$\frac{1 \text{ slice}}{\frac{1}{16} \text{ inch}} = \frac{n \text{ slices}}{2 \text{ inches}}$$

$$\frac{1}{16} \times n = 1 \times 2$$

$$n = 2 \div \frac{1}{16}$$

$$n = \frac{2}{1} \times \frac{16}{1} = \textbf{32 slices}$$

2. *necessary information:* 1,460 loaves, $1\frac{3}{4}$ teaspoons

labels for proportion: $\frac{\text{teaspoons}}{\text{loaves}}$

$$\frac{1\frac{3}{4} \text{ teaspoons}}{1 \text{ loaf}} = \frac{n \text{ teaspoons}}{1,460 \text{ loaves}}$$

$$1 \times n = 1\frac{3}{4} \times 1,460$$

$$n = \frac{7}{\cancel{4}_1} \times \cancel{1,460}^{365} = \textbf{2,555 teaspoons}$$

3. *necessary information:* $\frac{7}{8}$ inch, 35 inches

labels for proportion: $\frac{\text{inches}}{\text{books}}$

$$\frac{\frac{7}{8} \text{ inch}}{1 \text{ book}} = \frac{35 \text{ inches}}{n \text{ books}}$$

$$\frac{7}{8} \times n = 1 \times 35$$

$$n = 35 \div \frac{7}{8}$$

$$n = \cancel{35}^{5} \times \frac{8}{\cancel{7}_1} = \textbf{40 books}$$

4. *necessary information:* $9\frac{2}{3}$ ounces, 16 cans

labels for proportion: $\frac{\text{cans}}{\text{ounces}}$

$$\frac{1 \text{ can}}{9\frac{2}{3} \text{ ounces}} = \frac{16 \text{ cans}}{n \text{ ounces}}$$

$$1 \times n = 9\frac{2}{3} \times 16$$

$$n = \frac{29}{3} \times \frac{16}{1} = \frac{464}{3}$$

$$n = \textbf{154}\frac{2}{3} \textbf{ ounces}$$

5. *necessary information:* 8 cups, $\frac{1}{4}$ cup

labels for proportion: $\frac{\text{cups}}{\text{loads}}$

$$\frac{8 \text{ cups}}{n \text{ loads}} = \frac{\frac{1}{4} \text{ cup}}{1 \text{ load}}$$

$$\frac{1}{4} \times n = 8 \times 1$$

$$n = 8 \div \frac{1}{4}$$

$$n = \frac{8}{1} \times \frac{4}{1} = \textbf{32 loads}$$

Page 101

1. *conversion:* 12 months = 1 year

$$\frac{12 \text{ months}}{1 \text{ year}} = \frac{30 \text{ months}}{n \text{ years}}$$

$$12 \times n = 1 \times 30$$

$$12n = 30$$

$$n = \frac{30}{12} = \frac{5}{2} = \textbf{2}\frac{1}{2} \textbf{ years}$$

2. *conversion:* 4 quarts = 1 gallon

$$\frac{4 \text{ quarts}}{1 \text{ gallon}} = \frac{n \text{ quarts}}{200 \text{ gallons}}$$

$$1 \times n = 4 \times 200$$

$$n = 800 \text{ quarts} = \textbf{800 bottles}$$

3. *conversion:* 1,000 meters = 1 kilometer

$$\frac{1,000 \text{ meters}}{1 \text{ kilometer}} = \frac{10,000 \text{ meters}}{n \text{ kilometers}}$$

$$1,000 \times n = 1 \times 10,000$$

$$1,000n = 10,000$$

$$n = \frac{10,000}{1,000} = \textbf{10 kilometers}$$

4. *conversion:* 2,000 pounds = 1 ton

$$\frac{2,000 \text{ pounds}}{1 \text{ ton}} = \frac{n \text{ pounds}}{\frac{1}{2} \text{ ton}}$$

$$1 \times n = \frac{\cancel{2,000}^{1000}}{1} \times \frac{1}{\cancel{2}_1}$$

$$n = \textbf{1,000 pounds}$$

5. *conversion:* 32 ounces = 1 quart

$$\frac{32 \text{ ounces}}{1 \text{ quart}} = \frac{n \text{ ounces}}{3 \text{ quarts}}$$

$$1 \times n = 32 \times 3$$

$$n = \textbf{96 ounces}$$

6. *conversion:* 5,280 feet = 1 mile

$$\frac{5,280 \text{ feet}}{1 \text{ mile}} = \frac{29,028 \text{ feet}}{n \text{ miles}}$$

$$5,280 \times n = 1 \times 29,028$$

$$n = \frac{29,028}{5,280} = \textbf{5.5 miles}$$

Page 102

1. $\dfrac{\$12.60}{1 \text{ yard}} = \dfrac{\$n}{3\frac{1}{3} \text{ yards}}$

   $1 \times n = \$12.60 \times 3\frac{1}{3}$

   $n = \overset{4.20}{\cancel{12.60}} \times \dfrac{10}{\cancel{3}_1} = \$42.00$

2. $\dfrac{13.5 \text{ pounds}}{4\frac{1}{2} \text{ years}} = \dfrac{n \text{ pounds}}{1 \text{ year}}$

   $4\frac{1}{2} \times n = 13.5 \times 1$

   $\dfrac{9}{2}n = 13.5$

   $n = 13.5 \div \dfrac{9}{2}$

   $n = 13.5 \times \dfrac{2}{9} = \dfrac{27.0}{9} = \textbf{3 pounds}$

3. $\dfrac{2\frac{2}{3} \text{ pounds}}{\$3.25} = \dfrac{1 \text{ pound}}{\$n}$

   $2\frac{2}{3} \times n = 3.25 \times 1$

   $\dfrac{8}{3}n = 3.25$

   $n = 3.25 \div \dfrac{8}{3}$

   $n = 3.25 \times \dfrac{3}{8} = \dfrac{9.75}{8}$

   $n = 1.218 \text{ or } \textbf{\$1.22}$

4. $\dfrac{7\frac{1}{2} \text{ rolls}}{\$384.50} = \dfrac{1 \text{ roll}}{\$n}$

   $7\frac{1}{2} \times n = 1 \times 384.50$

   $\dfrac{15}{2}n = 384.50$

   $n = 384.50 \div \dfrac{15}{2}$

   $n = 384.50 \times \dfrac{2}{15} = \dfrac{769}{15}$

   $n = 51.266 \text{ or } \textbf{\$51.27}$

Page 103

1. *necessary information:* $\frac{1}{12}$ of an hour, 8 hours

   labels for proportion: $\dfrac{\text{chickens}}{\text{hours}}$

   $\dfrac{1 \text{ chicken}}{\frac{1}{12} \text{ hour}} = \dfrac{n \text{ chickens}}{8 \text{ hours}}$

   $\dfrac{1}{12} \times n = 1 \times 8$

   $n = 8 \div \dfrac{1}{12}$

   $n = 8 \times 12$

   $n = \textbf{96 chickens}$

2. *necessary information:* 8 blood samples, 60 minutes

   labels for proportion: $\dfrac{\text{minutes}}{\text{blood samples}}$

   $\dfrac{60 \text{ minutes}}{8 \text{ blood samples}} = \dfrac{n \text{ minutes}}{1 \text{ blood sample}}$

   $8 \times n = 1 \times 60$

   $n = \dfrac{60}{8} = \dfrac{15}{2}$

   $n = 7\frac{1}{2} \textbf{ minutes}$

3. *necessary information:* 1.6 kilometers, 26 miles

   labels for proportion: $\dfrac{\text{kilometers}}{\text{miles}}$

   $\dfrac{1.6 \text{ kilometers}}{1 \text{ mile}} = \dfrac{n \text{ kilometers}}{26 \text{ miles}}$

   $1 \times n = 1.6 \times 26$

   $n = \textbf{41.6 kilometers}$

4. *necessary information:* 6 feet, 24 hours

   labels for proportion: $\dfrac{\text{feet}}{\text{hours}}$

   $\dfrac{6 \text{ feet}}{1 \text{ hour}} = \dfrac{n \text{ feet}}{24 \text{ hours}}$

   $1 \times n = 6 \times 24$

   $n = \textbf{144 feet}$

5. *necessary information:* 0.04 ounce, 12-ounce can

   labels for proportion: $\dfrac{\text{grams}}{\text{ounces}}$

   $\dfrac{1 \text{ gram}}{0.04 \text{ ounce}} = \dfrac{n \text{ grams}}{12 \text{ ounces}}$

   $0.04 \times n = 1 \times 12$

   $0.04n = 12$

   $n = \dfrac{12}{0.04}$

   $n = \textbf{300 grams}$

6. *necessary information:* $3\frac{1}{4}$ pounds, 2 pumpkin pies, 10 pies

   labels for proportion: $\dfrac{\text{pounds}}{\text{pies}}$

   $\dfrac{3\frac{1}{4} \text{ pounds}}{2 \text{ pies}} = \dfrac{n \text{ pounds}}{10 \text{ pies}}$

   $2 \times n = 10 \times 3\frac{1}{4}$

   $2n = \dfrac{10}{1} \times \dfrac{13}{4}$

   $2n = \dfrac{130}{4}$

   $n = \dfrac{130}{4} \div 2$

   $n = \dfrac{\overset{65}{\cancel{130}}}{4} \times \dfrac{1}{\cancel{2}_1}$

   $n = \dfrac{65}{4}$

   $n = \textbf{16}\frac{1}{4} \textbf{ pounds}$

7. *necessary information:* 68,000 gallons of water

   labels for proportion: $\dfrac{\text{gallons}}{\text{hours}}$

   conversion: 1 day = 24 hours

   $\dfrac{68,000 \text{ gallons}}{1 \text{ hour}} = \dfrac{n \text{ gallons}}{24 \text{ hours}}$

   $1 \times n = 68,000 \times 24$

   $n = \textbf{1,632,000 gallons}$

8. *necessary information:* \$0.12, $4\frac{1}{4}$ feet

   labels for proportion: $\dfrac{\$}{\text{feet}}$

   $\dfrac{\$0.12}{1 \text{ foot}} = \dfrac{\$n}{4\frac{1}{4} \text{ feet}}$

   $1 \times n = \$0.12 \times 4\frac{1}{4}$

   $n = \overset{.03}{\cancel{\$0.12}} \times \dfrac{17}{\cancel{4}_1}$

   $n = \textbf{\$0.51}$

9. *necessary information:* $3\frac{1}{2}$ minutes

labels for proportion: $\frac{\text{seconds}}{\text{minutes}}$

$$\frac{60 \text{ seconds}}{1 \text{ minute}} = \frac{n \text{ seconds}}{3\frac{1}{2} \text{ minutes}}$$

$$1 \times n = 60 \times 3\frac{1}{2}$$

$$n = \overset{30}{\cancel{60}} \times \frac{7}{\underset{1}{\cancel{2}}}$$

$$n = \textbf{210 seconds}$$

10. *necessary information:* 942 pages, 302,382 words

labels for proportion: $\frac{\text{words}}{\text{pages}}$

$$\frac{302,382 \text{ words}}{942 \text{ pages}} = \frac{n \text{ words}}{1 \text{ page}}$$

$$942 \times n = 1 \times 302,382$$

$$942n = 302,382$$

$$n = \frac{302,382}{942}$$

$$n = \textbf{321 words}$$

Pages 104–105

1. multiplication	**6.** subtraction
2. division	**7.** division
3. addition	**8.** division
4. subtraction	**9.** addition
5. multiplication	

Pages 108–109

1. *necessary information labels:* ounce, pound
answer label: ounces
Are the labels different? yes
new label: ounces
ounces – ounces = ounces

$$\begin{array}{r} 20 \text{ pounds} = 320 \text{ ounces} \\ 320 \text{ ounces} \\ - \quad 8 \text{ ounces} \\ \hline \textbf{312 ounces} \end{array}$$

2. *necessary information labels:* feet, inches
answer label: inches OR feet
Are the labels different? yes
new label: inches OR feet
inches – inches = inches OR
feet – feet = feet
72 inches = 6 feet OR 2 feet = 24 inches

$$\begin{array}{cc} 6 \text{ feet} & 72 \text{ inches} \\ - 2 \text{ feet} & \text{OR} \quad - 24 \text{ inches} \\ \hline \textbf{4 feet} & \textbf{48 inches} \end{array}$$

3. *necessary information labels:* inches, inches
answer label: inches
Are the labels different? no
new label:
inches + inches = inches

$$\begin{array}{ccc} 9\frac{1}{2} \text{ inches} & = & 9\frac{2}{4} \text{ inches} \\ + 1\frac{3}{4} \text{ inches} & = & + 1\frac{3}{4} \text{ inches} \\ \hline & & 10\frac{5}{4} \text{ inches} = \textbf{11}\frac{1}{4} \textbf{ inches} \end{array}$$

4. $\frac{\text{sheets}}{\cancel{\text{reams}}} \times \cancel{\text{reams}} = \text{sheets}$

$\frac{500 \text{ sheets}}{1 \cancel{\text{ream}}} \times 10 \cancel{\text{reams}} = \textbf{5,000 sheets}$

5. dollars $\div \frac{\text{dollars}}{\text{tickets}} = \text{tickets}$

$\cancel{\text{dollars}} \times \frac{\text{tickets}}{\cancel{\text{dollars}}} = \text{tickets}$

$\$522 \div \frac{\$6}{\text{ticket}} = ? \text{ tickets}$

$\$522 \div 6 = \textbf{87 tickets}$

6. $\cancel{\text{plants}} \times \frac{\text{pounds}}{\cancel{\text{plant}}} = \text{pounds}$

$14 \cancel{\text{plants}} \times \frac{8 \text{ pounds}}{1 \cancel{\text{plant}}} = \textbf{112 pounds}$

Pages 110–114

1. b.	**4. c.**
2. d.	**5. d.**
3. a.	

6. e. current year – age = birth year
$1998 - 86 = \textbf{1912}$

7. d. meters per cable × number of cables = total meters
$35 \text{ meters} \times 7 \text{ cables} = \textbf{245 meters}$

8. c. old population + rise = new population
$782,250 \text{ people} + 37,250 \text{ people} = \textbf{819,500 people}$

9. a. $\frac{\text{watts}}{\text{lights}} = \frac{\text{watts}}{\text{lights}}$

$$\frac{2}{1} = \frac{300}{n}$$

$$2n = 300$$

$$n = \frac{300}{2} = \textbf{150 lights}$$

10. c. $\frac{\text{cables}}{\text{calls}} = \frac{\text{cables}}{\text{calls}}$

$$\frac{1}{12,500} = \frac{n}{87,500}$$

$$12,500n = 87,500$$

$$n = \frac{87,500}{12,500} = \textbf{7 cables}$$

11. e. not enough information given. You do not know the tax rate.

12. **a.** total cloth ÷ number of looms = cloth per loom
8,760 yards ÷ 60 looms = **146 yards of cloth**

13. **b.** month's sales – goal = extra sales
37 encyclopedias – 20 encyclopedias =
17 encyclopedias

Answers may vary.

14. How much more did the average person in the District of Columbia make than he or she spent?

15. By how much was the price reduced?

16. How much water must he add to the muriactic acid to treat a concrete floor?

17. How many square feet of ventilation did Rosalia need to put in her attic?

18. How much sand must he add to the cement to make a batch of base coat stucco?

19. With this setting, what is the largest size the drilled hole might be?

20. What was the total cost of staying overnight at Motel 5?

Pages 115–117

1. **b.** total bill – tax = cost of sandwich alone
$1.89 – $.09 = **$1.80**

2. **a.** United States – Russia = difference
260,714,000 people – 149,609,000 people =
111,105,000 people

3. **d.** string ÷ length of pieces = number of pieces
12 feet ÷ $\frac{3}{4}$-foot piece = **16 pieces**

4. **d.** total amount for chicken ÷ price per pound = weight of chicken breasts
$8.70 ÷ $1.95 = **4.46 pounds**

5. **b.** stock closing – stock opening = gain
$22\frac{1}{2} – 20\frac{3}{8} = 2\frac{1}{8}$

6. **c.** peaches + plums = total weight
1.72 pounds + 0.9 pound = **2.62 pounds**

7. **d.** servings × grams per serving = total grams
11 × 0.26 = **2.86 grams**

8. **b.** cost of repair – amount deductible = insurance payment
$1,125 – $250 = **$875**

9. **a.** fraction (of) × paycheck = food cost
$\frac{1}{3}$ × $414 = **$138**

10. **d.** total miles ÷ miles per check = total maintenance checks
96,000 miles ÷ 12,000 miles = **8 maintenance checks**

11. **a.** miles ÷ gallons = miles per gallon
283.1 miles ÷ 14.9 gallons = **19 miles per gallon**

12. **b.** normal depth + feet above normal = total depth
7 feet + 14 feet = **21 feet**

13. **c.** miles ÷ hours = average speed
3,855 ÷ $3\frac{3}{4}$ hours = **1,028 miles per hour**

Pages 119–120

1. percent	6. percent
2. part	7. part
3. whole	8. percent
4. part	9. part
5. whole	10. whole

Pages 124–126

1. *part:* 36
whole: 144
percent: n (is what %?)
$\frac{36}{144} = \frac{n}{100}$
$144 \times n = 36 \times 100$
$144n = 3,600$
$n = \frac{3,600}{144}$
$n = \textbf{25\%}$

2. *part:* 288
percent: 72%
whole: n (of what number?)
$\frac{288}{n} = \frac{72}{100}$
$72 \times n = 100 \times 288$
$72n = 28,800$
$n = \frac{28,800}{72}$
$n = \textbf{400}$

3. *whole:* 75
percent: 68%
part: n (What is?)
$\frac{n}{75} = \frac{68}{100}$
$100 \times n = 75 \times 68$
$100n = 5,100$
$n = \frac{5,100}{100}$
$n = \textbf{51}$

4. *whole:* $160
part: $40
percent: n (What was the percent?)
$\frac{40}{160} = \frac{n}{100}$
$160 \times n = 40 \times 100$
$160n = 4,000$
$n = \frac{4,000}{160}$
$n = \textbf{25\%}$

5. *percent:* 58%

whole: 28,450 votes

part: n (How many votes did Marsha receive?)

$$\frac{n}{28,450} = \frac{58}{100}$$

$$100 \times n = 58 \times 28,450$$

$$100n = 1,650,100$$

$$n = \frac{1,650,000}{100}$$

$$n = \textbf{16,501 votes}$$

6. *percent:* 25%

part: $96,000

whole: n (How much aid was Metropolis receiving?)

$$\frac{96,000}{n} = \frac{25}{100}$$

$$25 \times n = 96,000 \times 100$$

$$25n = 9,600,000$$

$$n = \frac{9,600,000}{25}$$

$$n = \textbf{\$384,000}$$

7. *percent:* 7%

part: $1,659

whole: n (What was his income?)

$$\frac{1,659}{n} = \frac{7}{100}$$

$$7 \times n = 1,659 \times 100$$

$$7n = 165,900$$

$$n = \frac{165,900}{7}$$

$$n = \textbf{\$23,700}$$

8. *percent:* 60%

whole: 345,780

part: n (How many African American people live in the city?)

$$\frac{n}{345,780} = \frac{60}{100}$$

$$100 \times n = 345,780 \times 60$$

$$100n = 20,746,800$$

$$n = \frac{20,746,800}{100}$$

$$n = \textbf{207,468 African American people}$$

9. *part:* $340

whole: $2,000

percent: n (What was the interest rate?)

$$\frac{340}{2,000} = \frac{n}{100}$$

$$2,000 \times n = 340 \times 100$$

$$2,000n = 34,000$$

$$n = \frac{34,000}{2,000}$$

$$n = \textbf{17\%}$$

10. *part:* n (What is the most calories from fat?)

whole: 2,400 calories

percent: 20%

$$\frac{n}{2,400} = \frac{20}{100}$$

$$100 \times n = 2,400 \times 20$$

$$100n = 48,000$$

$$n = \frac{48,000}{100}$$

$$n = \textbf{480 calories}$$

11. c. 20% is the amount she saved. You are looking for the dollar amount she saved.

12. e. 720 dentists recommended Never Break. You are looking for the percent that recommended Never Break.

13. a. 60% of the legislators have to support a tax increase in order for it to pass. You are looking for the number of legislators needed to support a tax increase.

14. b. You are given that $75 is 30% off list price. You are looking for the list price.

15. d. 2% is the smallest percent of defective parts needed to shut down the production line. You are looking for the smallest number of defective parts that would result in the production line being shut down.

Pages 127–128

1. b. Since a state has thousands of workers, the percent is much more likely to be an estimate.

2. a. Sale prices will be exact to the nearest penny.

3. b. Predicting a future event like an earthquake is always an estimation.

4. b. It is likely that the actual background radiation level would be a fraction of a percent more or less than 35%.

5. a. 88% on a 50-question multiple-choice test is precisely 44 questions correct.

6. b. Cities have populations of thousands or more. It is very likely that the number of people living in poverty is not exactly 24%.

7. Since the report is probably rounded to the nearest percent, she might have received anywhere from 3137 votes (64.5%) to 3186 votes (65.5%).

8. For the estimate to be accurate, the actual population must be between 134,210 and 135,495.

9. The actual percent was slightly over 11.7%, which is close enough to 12% for the prediction to be considered accurate.

10. If the state says 3% of all tickets are winners, it is likely that the number will be exactly accurate, even with the large number of tickets. 600,000 winning tickets were printed.

11. The increase was slightly over 23%. This was not accurate to the nearest percent, but for a business forecast was not that far off. Therefore, you could make a good argument either way.

Page 131

1. You are looking for the part. Therefore, you should multiply the total bill times the percent rate of the tip. $40.00 \times 0.15 =$ **$6.00**

2. You are looking for the part. Therefore, you should multiply the cost of the chain saw times the sales tax percent. $80.00 \times 0.05 =$ **$4.00**

3. You are looking for the percent. Therefore, you should divide the part, the employees over 60, by the whole, the total number of employees.

$$540\overline{)27.00} \quad \frac{.05}{}$$

Convert .05 to a percent: $.05 \times 100 =$ **5%**

4. You are looking for the whole. Therefore, you should divide the part, the customers who will visit the store, by the percent.

$$.02\overline{)1500.00} \quad \frac{75,000 \text{ people}}{}$$

5. You are looking for the percent. Therefore, you should divide the part, the graduates who found jobs, by the whole, the total number of graduates.

$$30\overline{)24.00} \quad \frac{.80}{}$$

Convert .80 to a percent: $.80 \times 100 =$ **80%**

6. You are looking for the percent. Therefore, you should divide the part, the amount spent, by the whole, the amount raised.

$$112000\overline{)28000.00} \quad \frac{.25}{}$$

Convert .25 to a percent: $.25 \times 100 =$ **25%**

Pages 133–134

1. *percent:* 4.5%
   part: 90
   whole: n (of what number?)
$$\frac{90}{n} = \frac{4.5}{100}$$
$$4.5 \times n = 90 \times 100$$
$$4.5n = 9,000$$
$$n = \frac{9,000}{4.5}$$
$$n = \mathbf{2,000}$$

2. *part:* $\frac{1}{10}$
   whole: $\frac{3}{4}$
   percent: n (is what percent?)
$$\frac{\frac{1}{10}}{\frac{3}{4}} = \frac{n}{100}$$
$$\frac{3}{4} \times n = 100 \times \frac{1}{10}$$
$$\frac{3}{4}n = 10$$
$$n = 10 \div \frac{3}{4}$$
$$n = 10 \times \frac{4}{3} = \frac{40}{3} = 13\frac{1}{3}$$
$$n = \mathbf{13\frac{1}{3}\%}$$

3. *percent:* $66\frac{2}{3}\%$
   part: 42
   whole: n (of what number?)
$$\frac{42}{n} = \frac{66\frac{2}{3}}{100}$$
$$66\frac{2}{3} \times n = 42 \times 100$$
$$66\frac{2}{3}n = 4,200$$
$$\frac{200n}{3} = 4,200$$
$$n = \overset{21}{\cancel{4,200}} \times \frac{3}{\cancel{200}_{1}} = 63$$
$$n = \mathbf{63}$$

4. *percent:* 6.4%
   whole: 800
   part: n (What is?)
$$\frac{n}{800} = \frac{6.4}{100}$$
$$100 \times n = 6.4 \times 800$$
$$100n = 5,120$$
$$n = \frac{5,120}{100}$$
$$n = \mathbf{51.2}$$

5. *percent:* 8%
   part: $49.76
   whole: n (How much money?)
$$\frac{49.76}{n} = \frac{8}{100}$$
$$8 \times n = 49.76 \times 100$$
$$8n = 4,976$$
$$n = \frac{4,976}{8}$$
$$n = \mathbf{\$622}$$

6. *whole:* $8.60
   percent: 5%
   part: n (How much was the tax?)
$$\frac{n}{8.60} = \frac{5}{100}$$
$$100 \times n = 8.60 \times 5$$
$$100n = 43$$
$$n = \frac{43}{100}$$
$$n = \mathbf{\$0.43}$$

7. *whole:* $42.50

percent: 6%

part: n (Find the amount of tax)

$$\frac{n}{42.50} = \frac{6}{100}$$

$$n \times 100 = 42.50 \times 6$$

$$100n = 255$$

$$n = \mathbf{\$2.55}$$

8. *whole:* $15.95

percent: 7%

part: n (Find the amount of tax)

$$\frac{n}{15.95} = \frac{7}{100}$$

$$n \times 100 = 15.95 \times 7$$

$$100n = 111.65$$

$$n = \mathbf{\$1.12}$$

9. *part:* $0.96

whole: $1.92

percent: n (By what percent?)

$$\frac{0.96}{1.92} = \frac{n}{100}$$

$$1.92 \times n = 0.96 \times 100$$

$$1.92n = 96$$

$$n = \frac{96}{1.92}$$

$$n = \mathbf{50\%}$$

10. *whole:* $12.80

percent: $12\frac{1}{2}$%

part: n (How much did Juan save?)

$$\frac{n}{12.80} = \frac{12\frac{1}{2}}{100}$$

$$100 \times n = 12.80 \times 12\frac{1}{2}$$

$$100n = \overset{6.40}{\cancel{12.80}} \times \frac{\overset{25}{\cancel{25}}}{\cancel{2}_1}$$

$$100n = 160$$

$$n = \frac{160}{100}$$

$$n = \mathbf{\$1.60}$$

11. *percent:* 3.5%

part: 12,000 calls resulting in new subscribers

whole: n (How many calls?)

$$\frac{12,000}{n} = \frac{3.5}{100}$$

$$3.5n = 1,200,000$$

$$n = \mathbf{342,857} \text{ (round to nearest call since you can't make a fraction of a call)}$$

Pages 135–136

1. *part:* 3 out of

whole: 4 dentists

percent: n (What percent of all dentists?)

$$\frac{3}{4} = \frac{n}{100}$$

$$4 \times n = 3 \times 100$$

$$4n = 300$$

$$n = \frac{300}{4} = \mathbf{75\%}$$

2. *percent:* 40%

part: 112,492 people

whole: n (How many registered voters?)

$$\frac{112,492}{n} = \frac{40}{100}$$

$$40 \times n = 112,492 \times 100$$

$$40n = 11,249,200$$

$$n = \frac{11,249,200}{40} = \mathbf{281,230 \text{ voters}}$$

3. *part:* 11,500 fewer visitors

whole: 34,500 visitors

percent: n (What was the percent drop?)

$$\frac{11,500}{34,500} = \frac{n}{100}$$

$$34,500 \times n = 11,500 \times 100$$

$$34,500n = 1,150,000$$

$$n = \frac{1,150,000}{34,500} = \mathbf{33\frac{1}{3}\%}$$

4. *whole:* $49,600,000

percent: 8.6%

part: n (How much money did the company make?)

$$\frac{n}{49,600,000} = \frac{8.6}{100}$$

$$100 \times n = 8.6 \times 49,600,000$$

$$100n = 426,560,000$$

$$n = \frac{426,560,000}{100} = \mathbf{\$4,265,600}$$

5. *part:* 0.4 ounce

percent: $16\frac{2}{3}$%

whole: n (What was the weight of its chocolate bar before the change?)

$$\frac{0.4}{n} = \frac{16\frac{2}{3}}{100}$$

$$16\frac{2}{3} \times n = 100 \times 0.4$$

$$16\frac{2}{3}n = 40$$

$$\frac{50}{3}n = 40$$

$$n = 40 \div \frac{50}{3}$$

$$n = \overset{4}{\cancel{40}} \times \frac{3}{\cancel{50}_5} = 4 \times \frac{3}{5} =$$

$$\frac{12}{5} = 2\frac{2}{5} \text{ or } \mathbf{2.4 \text{ ounces}}$$

6. *part:* 40

percent: 0.8%

whole: n (How many people over age 65?)

$$\frac{40}{n} = \frac{0.8}{100}$$

$$0.8 \times n = 40 \times 100$$

$$0.8n = 4,000$$

$$n = \frac{4,000}{0.8} = \mathbf{5,000 \text{ people over age 65}}$$

7. *percent:* 13%

whole: $11,694

part: n (How much did he pay?)

$$\frac{n}{11,694} = \frac{13}{100}$$

$$100 \times n = 13 \times 11,694$$

$$100n = 152,022$$

$$n = \frac{152,022}{100} = \mathbf{\$1,520.22}$$

8. *whole:* $17,548

percent: 7%

part: n (How much of a raise will he get?)

$$\frac{n}{17,548} = \frac{7}{100}$$

$$100 \times n = 17,548 \times 7$$

$$100n = 122,836$$

$$n = \frac{122,836}{100} = \textbf{\$1,228.36}$$

9. *part:* $18

percent: 9%

whole: n (What had her week's salary been?)

$$\frac{18}{n} = \frac{9}{100}$$

$$9 \times n = 18 \times 100$$

$$9n = 1,800$$

$$n = \frac{1,800}{9} = \textbf{\$200}$$

10. *part:* 506 scores

whole: 1,012 attempts

percent: n (What was his scoring percentage?)

$$\frac{506}{1,012} = \frac{n}{100}$$

$$1,012 \times n = 506 \times 100$$

$$1,012n = 50,600$$

$$n = \frac{50,600}{1,012} = \textbf{50\%}$$

11. *part:* $\frac{3}{4}$ cup

percent: 30%

whole: n (How many cups of cereal must you eat?)

$$\frac{\frac{3}{4}\, cup}{n} = \frac{30}{100}$$

$$\frac{3}{4} \times 100 = 30n$$

$$75 = 30n$$

$$\textbf{2}\frac{1}{2} \textbf{ cups} = n$$

12. *whole:* $3\frac{3}{4}$ ounces

percent: n (What percent of vitamin D is provided?)

part: 1 ounce

$$\frac{1\, ounce}{3\frac{3}{4}\, ounce} = \frac{n}{100}$$

$$1 \times 100 = 3\frac{3}{4} \times n$$

$$100 = \frac{15}{4}n$$

$$\textbf{26.7\% or } \textbf{26}\frac{2}{3}\textbf{\%} = n$$

Pages 139–141

Information given in the problem is written under the appropriate place in the word sentence or proportion. Your wording may differ slightly.

1. *solution sentence:*

earnings − money taken out = take-home pay
($380)

missing information:

taxes + union dues = money taken out
($149)   ($16)

2. *solution sentence:*

starting money − money spent = money left
($41)

missing information:

lunch + gas = money spent
($3)   ($22)

3. *solution sentence:*

total cost ÷ number of dolls = cost per doll
($720)

missing information:

boxes × dolls per box = number of dolls
(30)   (8)

4. *solution sentence:*

starting balance + transactions = new balance
($394)

missing information:

deposit − check = transactions
($201)   ($187)

5. *solution sentence:*

cost of blouses + cost of skirt = total spent
                    ($16)

missing information:

cost per blouse × number of blouses = cost of
($12)   (5)   blouses

6. *solution sentence:*

total amount ÷ monthly payments = amount of
                    (24)   each payment

missing information:

loan + interest = total amount to pay
($4,600) ($728)

7. *proportion:*

$$\frac{\text{numbers entered } (10,000)}{\text{minutes } (60)} = \frac{\text{numbers entered}}{\text{minutes } (15)}$$

8. *proportion:*

$$\frac{\text{grams per serving } (14)}{\text{saturated fat serving } (2)} = \frac{\text{grams per jar } (224)}{\text{saturated fat per jar}}$$

9. *solution sentence:*

$ separate − $ combination = savings
                    ($4.29)

missing information:

cheeseburger + fries + cola = $ separate
($1.89)   ($1.29) ($1.19)

10. *solution sentence:*

ward delegates + at large = total delegates

ward delegates + 5 = total delegates

missing information:

wards × delegates = ward delegates

8 × 4 = 32 ward delegates

solution:

32 + 5 = **37 delegates**

11. *solution sentence:*
customer price – cost = profit
163 – cost = profit
missing information:
parts + labor = cost
81 + 45 = $126 cost
solution:
163 – 126 = **$37 profit**

12. *solution sentence:*
raise – new expenses = net monthly increase
78 – new expenses = net monthly increase
missing information:
daily increase × number of days = new expenses
2 × 22 = $44 new expenses
solution:
78 – 44 = **$34 net monthly increase**

13. $\frac{\text{people}}{\text{pounds}} = \frac{\text{people}}{\text{pounds}}$

$$\frac{6}{3} = \frac{8}{n}$$
$6 \times n = 3 \times 8$
$6n = 24$
$n = \frac{24}{6} =$ **4 pounds**

14. $\frac{\text{last month \$}}{\text{last month kilowatts}} = \frac{\text{this month \$}}{\text{this month kilowatts}}$

$$\frac{60}{720} = \frac{n}{648}$$
$720 \times n = 60 \times 648$
$720n = 38{,}880$
$n =$ **$54**

15. *solution sentence:*
cubic feet per day × # of days = total cubic feet
cubic feet per day × # of days = 1,728 cubic feet
missing information:
cubic feet per load × loads per day = cubic feet per day
48 cubic feet × 6 loads per day = 288 cubic feet per day
solution:
288 cubic feet per day × n days = 1,728 cubic feet
$n = \frac{1{,}728}{288} =$ **6 work days**

Pages 143–144

1. *solution sentence:*
total cost ÷ total cookies = cost per cookie
14.40 ÷ total cookies = cost per cookie
missing information:
cookies × number of boxes = total cookies
20 × 6 = 120 cookies
solution:
14.40 ÷ 120 = **$0.12 per cookie**

2. *proportion:*
$\frac{\text{reduction}}{\text{total price}} = \frac{\text{percent}}{100}$

$$\frac{n}{400} = \frac{30}{100}$$
$100n = 12{,}000$
$n = 120$
original price – reduction = sale price
400 – 120 = **$280**

3. *solution sentence:*
pears + fruit cocktail = total fruit
pears + $17\frac{1}{2}$ = total fruit
missing information:
cans × contents = ounces of pears
$5 \times 9\frac{3}{4} = 48\frac{3}{4}$ ounces
solution:
$48\frac{3}{4} + 17\frac{1}{2} = \mathbf{66\frac{1}{4}}$ **ounces**

4. *solution sentence:*
total savings – cost = net savings
total savings – 2.49 = net savings
missing information:
savings × months = total savings
3.40 × 12 = $40.80
solution:
40.80 – 2.49 = **$38.31 net savings**

5. *solution sentence:*
original weight – loss = cooked weight
$\frac{1}{4}$ pound – loss = cooked weight
missing information:
weight × fraction = loss
$\frac{1}{4} \times \frac{1}{3} = \frac{1}{12}$ loss
solution:
$\frac{1}{4} - \frac{1}{12} = \frac{1}{6}$ **pound**

6. *solution sentence:*
dinner + tax = total cost
24 + tax = total cost
missing information:
dinner × percent = tax
24 × .06 = $1.44
solution:
24 + 1.44 = **$25.44 total**

7. *proportion:*
$\frac{\text{discount}}{\text{original price}} = \frac{\text{percent}}{100}$

$$\frac{n}{19.50} = \frac{30}{100}$$
$100n = 585$
$n = 5.85
original price – discount = sale price
19.50 – 5.85 = **$13.65**

8. *solution sentence:*
   payment ÷ months = monthly payment
   payment ÷ 12 = monthly payment
   missing information:
   amount – down payment = payment
   310.60 – 130 = $180.60
   solution:
   180.60 ÷ 12 = **$15.05 monthly payment**

9. *solution sentence:*
   price per bottle – cost per bottle = profit per bottle
   2.10 – cost per bottle = profit per bottle
   missing information:
   cost ÷ bottles = cost per bottle
   57 ÷ 30 = $1.90
   solution:
   2.10 – 1.90 = **$0.20 profit**

10. *proportion:*
   $$\frac{part}{whole} = \frac{percent}{100}$$
   $$\frac{profit}{57} = \frac{n\ percent}{100}$$
   missing information:
   bottles × profit per bottle = total profit
   30 × .20 = $6
   solution:
   $$\frac{6}{57} = \frac{n}{100}$$
   $$57 \times n = 6 \times 100$$
   $$57n = 600$$
   $$n = \frac{600}{57} = \textbf{10.5\%}$$

11. $$\frac{ounces\ salt}{ounces\ stew} = \frac{ounces\ salt}{ounces\ stew}$$
   $$\frac{0.2}{256} = \frac{n}{12}$$
   $$256 \times n = 0.2 \times 12$$
   $$256n = 2.4$$
   $$n = \textbf{.009 ounces salt}$$
   (or $n = .01$ rounded off)

12. *solution sentence:*
   total material ÷ material per outfit = # of outfits
   70 yards ÷ material per outfit = # of outfits
   missing information:
   blouse + shirt = material per outfit
   $\frac{2}{3} + 1\frac{1}{4} = 1\frac{11}{12}$ yard
   solution:
   70 yards ÷ $1\frac{11}{12}$ yard = **36 outfits**
   (discard remainder since you are asked for complete outfits)

Page 146

1. $20 - 9 = 11$

2. $20 - 6 + 3 = 14 + 3 = 17$

3. $157.8 - 5.7 - 1.1 = 152.1 - 1.1 = 151$

4. $19.2 + 7.8 = 27$

5. $10 \div 5 = 2$

6. $36 \div 3 + 6 = 12 + 6 = 18$

7. $36 \div 3 + 6 = 12 + 6 = 18$

8. $90 - 10 - 8 = 80 - 8 = 72$

9. $1.8 \div 0.02 - 2 = 90 - 2 = 88$

10. $56 - 24.5 = 31.5$

Page 148

1. **d.**

2. **a.**

3. **b.**

4. **c.**

5. **e.**

Pages 150–151

1. **b.**
   $46 + 48 = 94$
   $94 - 81 = $**$13**

2. **e.**
   $5 \times 256 = 1,280$
   $1,280 \div 8 = $**160 ounces**

3. **b.**
   $85 \times 80 = 6,800$
   $6,800 \div 100 = $**$68**

4. **b.**
   $8.68 \div 64 = 0.14$ (rounded to the nearest cent)
   $4.02 \div 32 = 0.13$ (rounded to the nearest cent)
   $0.14 - 0.13 = $**$.01 per ounce**

Page 153

1. **c.**

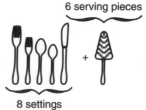

8 settings   6 serving pieces

2. e.
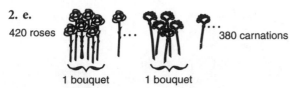
420 roses   380 carnations
1 bouquet   1 bouquet

3. d.
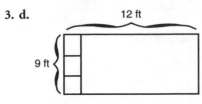
12 ft
9 ft

4. a.
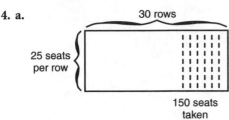
30 rows
25 seats per row
150 seats taken

Page 155

1. floor size ÷ tile size = number of tiles
   floor size ÷ 81 = number of tiles
   conversion:
   (144 square inches in a square foot)
   $$\frac{\text{square inches}}{1 \text{ square foot}} = \frac{n \text{ square inches}}{\text{square feet}}$$
   $$\frac{144}{1} = \frac{n}{54}$$
   $$n = 144 \times 54 = 7{,}776 \text{ square inches}$$
   solution:
   $7{,}776 \div 81 = $ **96 tiles**

2. welds per hour × hours = total welds
   $20 \times 9 = $ **180 total welds**

3. highway ÷ reflector distance = number of reflectors
   highway ÷ 528 = reflectors
   conversion: (1 mile = 5,280 feet)
   $$\frac{\text{feet}}{\text{mile}} = \frac{\text{feet}}{\text{mile}}$$
   $$\frac{5{,}280}{1} = \frac{n}{46}$$
   $$n = 242{,}880 \text{ feet}$$
   solution:
   $242{,}880 \div 528 = $ **460 reflectors**
   (Another acceptable answer would be 461 reflectors. This would depend on whether the first or the second reflector was at the first 528-foot mark.)

4. gallons needed × pints per gallon = total people
   gallons × 8 = total people
   (There are 8 pints in a gallon.)
   missing information:
   gallons – donations = gallons needed
   48 – 26 = 22 gallons needed
   solution:
   $22 \times 8 = $ **176 people**

5. total coal ÷ pounds per customer = customers
   total coal ÷ 400 = customers
   conversion: (There are 2,000 pounds in a ton.)
   $$\frac{\text{pounds}}{1 \text{ ton}} = \frac{\text{pounds}}{\text{ton}}$$
   $$\frac{2{,}000}{1} = \frac{n}{38}$$
   $$n = 76{,}000 \text{ pounds}$$
   solution:
   $76{,}000 \div 400 = $ **190 customers**

6. $$\frac{\text{numbers}}{\text{time}} = \frac{\text{numbers}}{\text{time}}$$
   (There are 60 minutes in an hour.)
   $$\frac{463}{5} = \frac{n}{60}$$
   $$5n = 27{,}780$$
   $$n = \frac{27{,}780}{5} = \textbf{5{,}556 numbers}$$

7. ice cream ÷ serving = number of people
   ice cream ÷ 4 = number of people
   conversion: (1 quart contains 32 ounces.)
   $$\frac{\text{quarts}}{\text{ounces}} = \frac{\text{quarts}}{\text{ounces}}$$
   $$\frac{1}{32} = \frac{12}{n}$$
   $$n = 384 \text{ ounces}$$
   solution:
   $384 \div 4 = $ **96 people**

Page 157

1. *necessary information:* 103,912 students,
   4,657 students, 1,288 students
   original students – change = new total
   103,912 – change = new total
   missing information:
   left – enrolled = change
   4,657 – 1,288 = 3,369 students
   solution:
   103,912 – 3,369 = **100,543 students**

2. *necessary information:* 103,912 students,
   1,288 students
   original + new enrollees = total students
   103,912 + 1,288 = **105,200 students**

3. *necessary information:* 54 degrees, 27 degrees, 19 degrees
   high temperature – drop = midnight temperature
   high temperature – 19 = midnight temperature
   missing information:
   first + rise = high temperature
   54 + 27 = 81 degree high
   solution:
   81 – 19 = **62 degrees midnight temperature**

4. *necessary information:* $348, 52 weeks
   weekly income × number of weeks = yearly income
   348 × 52 = **$18,096**

5. *necessary information:* 3 paperbacks, 2 magazines, $6.95, $3.50, $50
   money paid – cost of purchases = change
   $50 – cost of purchases = change
   missing information: price of books + price of magazines = cost of purchases
   6.95 × 3 + 3.50 × 2 = $27.85
   solution:
   50.00 – 27.85 = **$22.15**

6. *necessary information:* 20%, $2.80
   cost – amount of discount = final price
   $2.80 – amount of discount = final price
   missing information:
   cost × percent = amount of discount
   2.80 × 0.20 = $0.56
   solution:
   2.80 – 0.56 = **$2.24 final price**

Page 159

1. apples + cantaloupe = total
   apples + 1.88 = total
   missing information:
   $\frac{cost}{apples} = \frac{cost}{apples}$
   $\frac{2.76}{12} = \frac{n}{7}$
   $12n = 19.32$
   $n = 19.32 \div 12 = \$1.61$
   solution:
   1.61 + 1.88 = **$3.49**

2. bill – discount = new bill
   36.80 – discount = new bill
   missing information:
   bill × percent = discount
   36.80 × 0.06 = 2.208 = $2.21
   solution:
   36.80 – 2.21 = **$34.59**

3. reduced calories – breakfast = rest of day
   reduced calories – 797 = rest of day
   missing information:
   original – (percent × calories) = reduced calories
   4,200 – (0.28 × 4,200) = reduced calories
   4,200 – 1,176 = 3,024 calories
   solution:
   3,024 – 797 = **2,227 calories**

4. salary + commission = pay
   70 + commission = pay
   missing information:
   percent × sales over 200 = commission
   0.06 × (4,160 – 200) = commission
   0.06 × 3,960 = $237.60
   solution:
   70 + 237.60 = **$307.60**

5. money spent ÷ gallons = price per gallon
   186 ÷ gallons = price per gallon
   missing information:
   miles ÷ miles per gallon = gallons
   3,627 ÷ 31 = 117 gallons
   solution:
   186 ÷ 117 = 1.589 or **$1.59 per gallon**

Pages 160–161

1. **c.** daily miles × days = total miles
   daily miles × 5 = total
   missing information:
   to work + back + delivery = daily miles
   7 + 7 + 296 = 310 miles
   solution:
   310 × 5 = **1,550 miles**

2. **a.** travel miles + miles at work = daily miles
   14 + 296 = 310 miles

3. **c.** total ÷ students = individual cost
   total ÷ 15 = individual cost
   missing information:
   books + materials = total
   135 + 225 = $360
   solution:
   360 ÷ 15 = **$24 individual cost**

4. **e.** small rooms' capacities + main room capacity = total capacity
   small rooms' capacities + 94 = total capacity
   missing information:
   rooms × capacity = small rooms' capacities
   4 × 28 = 112 people
   solution:
   112 + 94 = **206 people**

5. b. players × teams = players before change
36 × 8 = **288 players** before change

6. d. players on new rosters × teams = total after change
players on new rosters × 8 = total after change
missing information:
players − reduction = players on new rosters
36 − 3 = 33 players
solution:
33 × 8 = **264 players**

7. d. cheese cost + apples cost = total cost
missing information:
pounds × cost = cheese cost
2.36 × 2.58 = $6.0888 or $6.09
pounds × cost = apples cost
4 × 0.89 = $3.56
solution:
6.09 + 3.56 = **$9.65**

8. c. front-loading − top-loading = additional loads
missing information:
cups ÷ amount = front-loading
$6 \div \frac{1}{4} = 6 \times 4 = 24$ loads
cups ÷ amount = top-loading
$6 \div \frac{1}{3} = 6 \times 3 = 18$ loads
solution:
24 − 18 = **6 loads**

Pages 162–165, Posttest A

1. $\dfrac{\text{cocoa}}{\text{chocolate}} = \dfrac{\text{cocoa}}{\text{chocolate}}$

$\dfrac{3 \text{ tablespoons}}{1 \text{ ounce}} = \dfrac{n \text{ tablespoons}}{12 \text{ ounces}}$

$1 \times n = 3 \times 12$
$n = $ **36 tablespoons**

2. $\dfrac{\text{part}}{\text{whole}} = \dfrac{\text{percent}}{100}$

$\dfrac{25 \text{ square inches}}{400 \text{ square inches}} = \dfrac{n\,\%}{100}$

$400 \times n = 25 \times 100$
$n = 2,500 \div 400 = 6\frac{1}{4}\%$

3. total children − children not in public school = children in public school
5,372 children − 1,547 children = **3,825 children**

4. $\dfrac{\text{pounds}}{\text{bushel}} = \dfrac{\text{pounds}}{\text{bushel}}$

$\dfrac{48 \text{ pounds}}{1 \text{ bushel}} = \dfrac{12 \text{ pounds}}{n \text{ bushels}}$

$48 \times n = 1 \times 12$
$n = 12 \div 48 = \frac{1}{4}$ **bushel**

5. hot dog + soda = total spent
$1.30 + $0.65 = **$1.95**

6. $\dfrac{\text{part}}{\text{whole}} = \dfrac{\text{percent}}{100}$

missing information: 80% − 70% = 10%

$\dfrac{n}{28,657} = \dfrac{10}{100}$

$100 \times n = 10 \times \$28,657$
$n = 286,570 \div 100 = $ **$2,865.70**

7. $\dfrac{\text{liquid}}{\text{corn syrup}} = \dfrac{\text{liquid}}{\text{corn syrup}}$

$\dfrac{\frac{1}{4} \text{ cup liquid}}{1 \text{ cup corn syrup}} = \dfrac{n \text{ cup liquid}}{1\frac{1}{2} \text{ cups corn syrup}}$

$1 \times n = \frac{1}{4} \times 1\frac{1}{2}$
$n = \frac{1}{4} \times \frac{3}{2} = \frac{3}{8}$ **cup**

8. charge per member × number of members = total collected
missing information: dues + magazine = charge per member
20 + 15 = $35
35 × 13,819 members = **$483,665**

9. $\dfrac{\text{part}}{\text{whole}} = \dfrac{\text{percent}}{100}$

$\dfrac{\$16,000}{n} = \dfrac{8\%}{100}$

$8 \times n = 16,000 \times 100$
$n = 1,600,000 \div 8 = $ **$200,000**

10. P. original + N.Y. original + G. original = total value
1,346 + 658 + 4 = **$2,008**

11. original value − value lost = 5-year value
missing information:
$\frac{1}{3}$ original value + $\frac{1}{4}$ original value = value lost
$\frac{1}{3} \times 9,600 + \frac{1}{4} \times 9,600 =$ value lost
3,200 + 2,400 = $5,600
9,600 − 5,600 = **$4,000**

12. feet in a mile × number of miles = total number of feet
missing information:
feet in a yard × yards in a mile = feet in a mile
3 feet × 1,760 yards = 5,280 feet
5,280 feet × 5 miles = **26,400 feet**

13. original weight − weight lost = new weight
$172\frac{1}{2}$ pounds − $47\frac{3}{4}$ pounds = $124\frac{3}{4}$ **pounds**

14. *missing information:* You need to know how many miles are in a lap.

15. profit ÷ number of women = profit per woman
missing information:
earnings − expenses = profit
336,460 − 123,188 = $213,272
213,272 ÷ 4 women = **$53,318**

16. $\dfrac{\text{part}}{\text{whole}} = \dfrac{\text{percent}}{100}$

$\dfrac{n}{260} = \dfrac{20}{100}$

$100 \times n = 20 \times 260$

$n = 5{,}200 \div 100 = \textbf{\$52}$

17. voters for incumbent + voters for challenger = decided voters
$\dfrac{1}{3} + \dfrac{1}{4} = \dfrac{7}{12}$ **of the voters**

18. \$ per hour × number of hours = total earned
$\$7.20 \times 17.5 \text{ hours} = \textbf{\$126.00}$

19. *missing information:* You need to know the base price of the car.

20. total homes ÷ number of people = homes per person
948 homes ÷ 12 people = **79 homes**

21. total bill − air-conditioning = bill without air conditioner
$86.29 - 59.00 = \textbf{\$27.29}$

22. cost of socks + cost of towels = total spent
missing information:
cost per sock × number of socks = cost of socks
$\$1.79 \times 3 \text{ socks} = \$5.37$
cost per towel × number of towels = cost of towels
$\$2.69 \times 4 \text{ towels} = \$10.76$
$\$5.37 + \$10.76 = \textbf{\$16.13}$

23. container size ÷ serving size = number of servings
$9 \text{ pounds} \div \dfrac{1}{16} = \textbf{144 sundaes}$

24. dinner + movie + baby-sitter = cost of evening
$\$24.43 + \$7.50 + \$20.00 = \textbf{\$51.93}$

25. total weight ÷ weight per book = number of books
34.2 pounds ÷ 0.6 pound = **57 books**

Pages 167–172, Posttest B

1. c. conversion: 1 dozen eggs = 12 eggs

$\dfrac{1\frac{1}{2} \text{ pounds}}{12 \text{ eggs}} = \dfrac{n \text{ pounds}}{8 \text{ eggs}}$

$12 \times n = 1\frac{1}{2} \times 8$

$n = \dfrac{3}{2} \times 8 \times \dfrac{1}{12} = \dfrac{24}{24} = \textbf{1 pound}$

2. c. original cost per tool set − price of last set = money lost on last set
missing information:
total cost for sets ÷ number of sets = original cost per tool set
540.60 ÷ 15 tool sets = 36.04
$36.04 - 24.00 = \textbf{\$12.04}$

3. c. weekend's hot dogs − Saturday's = Sunday's hot dogs
426 hot dogs − 198 hot dogs = **228 hot dogs**

4. b. $\dfrac{\text{part}}{\text{whole}} = \dfrac{\text{percent}}{100}$

$\dfrac{156{,}000}{n} = \dfrac{39}{100}$

$39 \times n = 156{,}000 \times 100$

$n = 15{,}600{,}000 \div 39 = \textbf{400{,}000 votes}$

5. b. passengers per jet × number of jets = total passengers
214 passengers × 96 jets = **20,544 passengers**

6. b. tile − size needed = size cut off
$\dfrac{3}{4} \text{ foot} - \dfrac{1}{3} \text{ foot} = \dfrac{5}{12} \textbf{ foot}$

7. e. You need to know how many apples were in the bushel.

8. c. $\dfrac{\text{concrete}}{\text{square feet}} = \dfrac{\text{concrete}}{\text{square feet}}$

$\dfrac{1.23 \text{ cubic yards}}{100 \text{ square feet}} = \dfrac{n \text{ cubic yards}}{550 \text{ square feet}}$

$100 \times n = 1.23 \times 550$

$n = 676.50 \div 100 = 6.765 \text{ cubic yards}$

$n = \textbf{6.77 cubic yards}$

9. c. pounds for recipe − pounds in freezer = pounds needed
4 pounds − 2.64 pounds = **1.36 pounds**

10. c. $\dfrac{\text{gallons}}{\text{people}} = \dfrac{\text{gallons}}{\text{people}}$

$\dfrac{1{,}638{,}000 \text{ gallons}}{78{,}000 \text{ people}} = \dfrac{n \text{ gallons}}{1 \text{ person}}$

$n \times 78{,}000 = 1 \times 1{,}638{,}000$

$n = 1{,}638{,}000 \div 78{,}000 = \textbf{21 gallons}$

11. b. total weekly earnings × number of weeks = gross yearly earnings
missing information:
money taken out + take-home pay = total weekly earnings
$98.23 + 332.77 = \$431.00$
$431.00 \times 52 \text{ weeks} = \textbf{\$22{,}412.00}$

12. a. closed stations + remaining stations = last year's stations
423 stations + 2,135 stations = **2,558 service stations**

13. a. fraction (of) × total representatives = votes
$\dfrac{2}{3} \times 435 = \textbf{290 votes}$

14. e. You need to know how much she deposited.

15. b. inches for door + inches for quarter panel = inches needed
$33\frac{3}{4} \text{ inches} + 51\frac{2}{3} \text{ inches} = \textbf{85}\frac{5}{12} \textbf{ inches}$

16. d. weight of roast × price per pound = total price
2.64 pounds × 3.96 = $10.4544 = **$10.45**

17. c. $\frac{432 \text{ male workers}}{\text{total work force}} = \frac{\text{percent male workers}}{100}$

missing information:
100% work force − percent female = percent male
100% − 28% = 72 male
$\frac{432 \text{ male workers}}{n \text{ workers}} = \frac{72}{100}$
$72 \times n = 432 \times 100$
$n = 43{,}200 \div 72 = \textbf{600 workers}$

18. c. weight of carton ÷ number of nails = weight per nail
$\frac{3}{4}$ pound ÷ 75 nails = $\frac{3}{4} \times \frac{1}{75} = \frac{1}{100}$ pound = **0.01 pound**

19. a. $\frac{\text{part}}{\text{whole}} = \frac{\text{percent}}{100}$

$\frac{7 \text{ field goals}}{25 \text{ field goal attempts}} = \frac{n}{100}$
$25 \times n = 7 \times 100$
$n = 700 \div 25 = \textbf{28\%}$

20. d. original weight + first month + second month = new weight
104 pounds + 3 pounds + 4 pounds = **111 pounds**

21. a. roast + steak = total meat
3.69 pounds + 1.23 pounds = **4.92 pounds**

22. b. original balance − total checks + deposit = new balance
missing information:
first check + second check = total checks
46.19 + 22.45 = 68.64
74.81 − 68.64 + 60.00 = **$66.17**

23. d. $\frac{\text{part}}{\text{whole}} = \frac{\text{percent}}{100}$
$\frac{n \text{ pounds}}{450 \text{ pounds}} = \frac{9}{100}$
$100 \times n = 9 \times 450$
$n = 4{,}050 \div 100 = \textbf{40.5 pounds}$

24. b. $\frac{\text{miles}}{\text{gallon}} = \frac{\text{miles}}{\text{gallon}}$

$\frac{168 \text{ miles}}{5.6 \text{ gallons}} = \frac{417 \text{ miles}}{n \text{ gallons}}$
$168 \times n = 5.6 \times 417$
$n = 2{,}335.2 \div 168 = \textbf{13.9 gallons}$

25. e. $\frac{\$}{\text{pound}} = \frac{\$}{\text{pound}}$

$\frac{\$1.79}{1 \text{ pound}} = \frac{\$1.06}{n \text{ pound}}$
$1.79 \times n = 1 \times 1.06$
$n = 1.06 \div 1.79 = \textbf{0.59 pound}$

Pages 177–179

1. c. grams carbs OJ + grams carbs milk shake + grams carbs cola = total grams carbs beverages
22 grams + 60 grams + 34 grams = **116 grams carbohydrates**

2. d. proportion:
$\frac{12 \text{ ounces shake}}{10 \text{ grams fat}} = \frac{18 \text{ ounces shake}}{n \text{ grams fat}}$
$12n = 180$
$n = \textbf{15 grams fat}$

3. d. calories steak − calories salmon = difference in calories
914 calories − n calories = difference in calories
proportion:
$\frac{8 \text{ ounces}}{412 \text{ calories}} = \frac{12 \text{ ounces}}{n \text{ calories}}$
$8n = 4{,}944$
$n = 618 \text{ calories}$
914 calories − 618 calories = **296 calories**

4. Answers may vary. Make reasonable estimations.
Steak:
$\frac{12 \text{ ounces}}{914 \text{ calories}} = \frac{n \text{ ounces}}{600 \text{ calories}}$
$900n = 7{,}200$ (914 rounded to 900)
$n = \textbf{8 ounces}$

Hamburger on bun: $\frac{5 \text{ ounces}}{405 \text{ calories}} = \frac{n \text{ ounces}}{600 \text{ calories}}$
$400n = 3{,}000$ (405 rounded to 400)
$n = \textbf{7.5 ounces}$

Fried chicken breast: $\frac{6 \text{ ounces}}{442 \text{ calories}} = \frac{n \text{ ounces}}{600 \text{ calories}}$
$450n = 3{,}600$ (442 rounded to 450)
$n = \textbf{8 ounces}$

Spaghetti with tomato sauce:
$\frac{10 \text{ ounces}}{246 \text{ calories}} = \frac{n \text{ ounces}}{600 \text{ calories}}$
$250n = 6{,}000$ (246 rounded to 250)
$n = \textbf{24 ounces}$

Broiled salmon: $\frac{8 \text{ ounces}}{412 \text{ calories}} = \frac{n \text{ ounces}}{600 \text{ calories}}$
$400n = 4{,}800$ (412 rounded to 400)
$n = \textbf{12 ounces}$

5. fat from 10-ounce sirloin steak – fat from other
entree = how much less fat

$$\frac{12\ \text{ounces}}{57\ \text{grams fat}} = \frac{10\ \text{ounces}}{n\ \text{grams fat}}$$

$$12n = 570$$
$$n = \textbf{47.5 grams fat from 10-ounce sirloin steak}$$

hamburger on bun: $\frac{5\ \text{ounce}}{21\ \text{grams fat}} = \frac{10\ \text{ounce}}{n\ \text{grams fat}}$
$$5n = 210$$
$$n = 42\ \text{grams fat}$$
$$47.5 - 42 = \textbf{5.5 grams less fat}$$

fried chicken breast: $\frac{6\ \text{ounces}}{22\ \text{grams fat}} = \frac{10\ \text{ounces}}{n\ \text{grams fat}}$
$$6n = 220$$
$$n = 36.7\ \text{grams from fat (rounded to nearest tenth)}$$
$$47.5 - 36.7 = \textbf{10.8 grams less fat}$$

spaghetti with tomato sauce: $47.5 - 1 = \textbf{46.5 grams less fat}$

salmon: $\frac{8\ \text{ounces}}{17\ \text{grams fat}} = \frac{10\ \text{ounces}}{n\ \text{grams fat}}$
$$8n = 170$$
$$n = 21.25$$
$$47.5 - 21.25 = \textbf{26.25 grams less fat}$$

6. Hamburger is ground up steak. According to the chart, steak has no carbohydrates. Therefore, the carbohydrates in a hamburger on bun come from the bun.

7. e. Broiled salmon gives the best combination of high protein and low fat. The steak has more protein but far more fat. The spaghetti with tomato sauce has much less fat but also much less protein. The hamburger on bun and the fried chicken have both less protein and more fat than the salmon.

8. c. and **e.**
jogging for 40 min + swimming for 30 min + bicycling for 2 hours (10 cal/min × 40 min) + (4 cal/min × 30 min) + (4 cal/min × 120 min)
400 cal + 120 cal + 480 cal = **1000 calories**

cross-country skiing for 2 hours + aerobic dancing for 35 min
(6 cal/min × 120 min) + (8 cal/min × 35 min)
720 cal + 280 cal = **1000 calories**

9. b. and **d.**
running for 25 min + walking for 45 min + bicycling for 25 min
(19 cal/min × 25 min) + (5 cal/min × 45 min) + (4 cal/min × 25 min)
475 cal + 225 cal + 100 cal = **800 calories**

aerobic dancing for 40 min + tennis for 40 min + jogging for 24 min
(8 cal/min × 40 min) + (6 cal/min × 40 min) + (10 cal/min × 24 min)
320 cal + 240 cal + 240 cal = **800 calories**

10. a.
recreational volleyball for 90 min + tennis for 55 min
(3 cal/min × 90 min) + (6 cal/min × 55 min)
270 cal + 330 cal = **600 calories**

11. Answers will vary. Make sure that total minutes are at least 60.
walking for 100 minutes
5 cal/min × 100 min = 500 calories

swimming for 40 minutes + jogging for 34 minutes
(4 cal/min × 40 min) + (10 cal/min × 34 min)
160 cal + 340 cal = 500 calories

12. Answers will vary. Make sure that total minutes is exactly 60.
aerobic dancing for 30 minutes + swimming for 30 minutes
(8 cal/min × 30 min) × (4 cal/min × 30 min)
240 cal + 120 cal = 360 calories

cross-country skiing for 40 minutes + walking for 20 minutes
(6 cal/min × 40 min) + (5 cal/min × 20 min)
240 cal + 100 cal = 340 calories

Pages 181–182

1. $\frac{200\ \text{cubic feet of water}}{1\ \text{hour}} = \frac{\text{total cubic feet of water}}{n\ \text{hours}}$

$$8\ \text{inches} = \frac{2}{3}\ \text{foot}$$

$$30\ \text{feet} \times 40\ \text{feet} \times \frac{2}{3}\ \text{feet} = \text{total cubic feet of water}$$
$$800\ \text{cubic feet} = \text{total cubic feet of water}$$

$$\frac{200\ \text{cubic feet of water}}{1\ \text{hour}} = \frac{800\ \text{cubic feet of water}}{n\ \text{hours}}$$
$$200n = 800$$
$$n = \textbf{4 hours}$$

2. water pumped out − water in basement = water seeped in

water pumped out = 300 cubic feet per hour × 6 hours = 1800 cubic feet

9 inches = $\frac{3}{4}$ foot

water in basement = 30 feet × 50 feet × $\frac{3}{4}$ foot = 1125 cubic feet

1800 cubic feet − 1125 cubic feet = **675 cubic feet**

3. The shortest route is **4.8 miles.** One way to get this total is to go to the gas station, then the video store, the department store, the library, the grocery, and finally home.

4. $\frac{\text{number of min}}{\text{change in sec}} = \frac{\text{number of min}}{\text{change in sec}}$

$\frac{10 \text{ min}}{5 \text{ sec change}} = \frac{n \text{ min}}{10 \text{ sec change}}$

$5n = 100$

$n = $ **20 minutes**

5. See if you can discover why a square would have the smallest possible perimeter.

$A = lw = l \times l$

$400 = l \times l$

$l = $ **20 feet.** The dimensions of the garden should be 20 feet.

6. $P = 2l + 2w$

$100 = 2l + 2(3)$

$100 = 2l + 6$

$94 = 2l$

47 feet $= l$

The run could be 47 feet long.

7. convert miles per hour to feet per second

$\frac{95 \text{ miles}}{1 \text{ hour}} \times \frac{5,280 \text{ ft}}{1 \text{ mile}} = \frac{501,600 \text{ ft}}{1 \text{ hour}}$

$\frac{501,600 \text{ ft}}{1 \text{ hour}} \times \frac{1 \text{ hour}}{60 \text{ min}} = \frac{8,360 \text{ ft}}{1 \text{ min}}$

$\frac{8,360 \text{ ft}}{1 \text{ min}} \times \frac{1 \text{ min}}{60 \text{ sec}} = \frac{139 \text{ ft}}{1 \text{ sec}}$

$\frac{139 \text{ ft}}{1 \text{ sec}} = \frac{60.5 \text{ ft}}{n \text{ sec}}$

proportion: $\frac{139 \text{ ft}}{1 \text{ sec}} = \frac{60.5 \text{ ft}}{n \text{ sec}}$

$139n = 60.5$

$n = \frac{60.5}{139} = $ **.44 sec**

(rounded to the nearest hundredth)

Using a Calculator

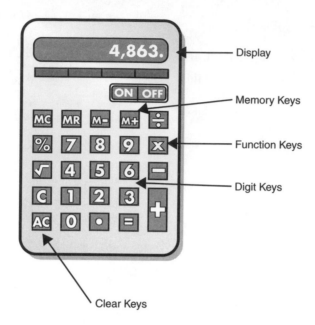

Display

Memory Keys

Function Keys

Digit Keys

Clear Keys

The calculator is an inexpensive convenient tool that you can use to help you with computation. Every calculator has a numeric keypad, keys for the four operations (+ for addition, – for subtraction, × for multiplication, and ÷ or / for division), an equal (=) key, and a clear (C) key. Most calculators also have keys for clear all (AC), percent (%), and square root (√). They may also have keys for the memory functions: memory add (M+), memory subtract (M–), memory read (MR), and memory clear (MC).

If you have never used a calculator before, you should begin by using it to add, subtract, multiply, and divide. Before you use a calculator, you should estimate what you expect your answer to be. If you do a good job of estimating, you should be able to recognize errors caused by accidentally hitting the wrong key on the calculator.

To add, subtract, multiply, or divide two numbers, follow this procedure.
Key in the first number.
Key in the operation (+, –, ×, ÷).
Key in the second number.
Press the equal sign (=).
The answer should appear in the display.

EXAMPLE   Multiply 48 × 396.

Key in 48 (First press 4 and then press 8.) [4] [8]

Key in ×. [×]

Key in 396 (First press 3, then press 9, and then press 6.) [3] [9] [6]

Press =. [=]

Look at the display. It should read 19,008. [19,008.]

Using Mental Math

Once you decide whether you need to add, subtract, multiply, or divide to solve a word problem, see if you might be able to solve the problem using mental math, or doing the math in your head. If you know your math facts, you should be able to solve a number of problems in this book, as well as many math problems you might encounter in your daily life, by using mental math. Throughout this book, you'll see the mental math icon for those problems you should try to solve using mental math.

The key to mental math is the basic addition and multiplication facts in these tables.

+	1	2	3	4	5	6	7	8	9
1	2	3	4	5	6	7	8	9	10
2	3	4	5	6	7	8	9	10	11
3	4	5	6	7	8	9	10	11	12
4	5	6	7	8	9	10	11	12	13
5	6	7	8	9	10	11	12	13	14
6	7	8	9	10	11	12	13	14	15
7	8	9	10	11	12	13	14	15	16
8	9	10	11	12	13	14	15	16	17
9	10	11	12	13	14	15	16	17	18

×	1	2	3	4	5	6	7	8	9
1	1	2	3	4	5	6	7	8	9
2	2	4	6	8	10	12	14	16	18
3	3	6	9	12	15	18	21	24	27
4	4	8	12	16	20	24	28	32	36
5	5	10	15	20	25	30	35	40	45
6	6	12	18	24	30	36	42	48	54
7	7	14	21	28	35	42	49	56	63
8	8	16	24	32	40	48	56	64	72
9	9	18	27	36	45	54	63	72	81

For example, you can use the table to find 3 × 4. You can use the same table to find 12 ÷ 3.

×	1	2	③	4
1	1	2	3	4
2	2	4	6	8
3	3	6	9	12
④	4	8	⑫	16
5	5	10	15	20

The more you use these math arithmetic facts, the more likely you will be able to memorize them. You could also use flash cards to help with your memorization.

Keep these tables face down when you are doing calculations. If you are not sure of a math fact, write down a guess and turn over the table to see how close your guess was. Using mental math is a skill that you can improve if you are willing to try to use it often.

Using Estimation

An estimate is an approximate answer. Estimation is one of the best ways to figure out whether or not your calculations make any sense. You can estimate either before or after you do a calculation. Many times a good estimate is accurate enough for your purposes and you will have no need to do an exact calculation at all.

Use Common Sense

Use your own knowledge and common sense to make estimates. If you fill a shopping cart full of groceries, will your bill be about $1, $10, $100, or $1,000? Without even knowing what items were purchased, you should be able to estimate that the cart full of groceries would cost closer to $100 than the other choices.

Use Rounded Numbers

When you are given numbers in a word problem, use rounded numbers to make an estimate. For example, gasoline costs $1.199 per gallon and you fill your gas tank with 15.121 gallons. You can round $1.199 to $1.20 and round 15.121 to 15. Then estimate the total cost by multiplying $1.20 by 15 to get an estimate of $18. You would expect the total cost of gasoline to cost about $18.

Use Friendly Numbers

You will often have situations in real life where exact math is not necessary. If you know your basic math facts, you can estimate by using friendly numbers—numbers that are close to the real numbers but that come out evenly. You will not get an exact answer, but you will get a result that is close enough for your purposes. For example, suppose you are in a grocery store and want to get the best buy on cereal. A 12-ounce box costs $2.29, while an 18-ounce box costs $2.99. If you round $2.29 to $2.40 (because 24 can be divided easily by 12), you can see that the 12-ounce box costs a little less than 20 cents an ounce. If you multiply 20 cents × 18 ounces, you get $3.60, much more than the $2.99 cost of the 18-ounce box. You can determine that the 18-ounce box is the better buy.

Develop your estimation skills by first estimating and then doing the exact calculations. The goal in estimating is not to be exact but to be close enough for your needs.

Formulas

PERIMETER

Figure	Name	Formula	Meaning
w / l	Rectangle	$P = 2l + 2w$	l = length w = width
s	Square	$P = 4s$	s = side

AREA

Figure	Name	Formula	Meaning
w / l	Rectangle	$A = lw$	l = length w = width
s	Square	$A = s^2$	s = side

VOLUME

Figure	Name	Formula	Meaning
h / w / l	Rectangular solid	$V = lwh$	l = length w = width h = height
s	Cube	$V = s^3$	s = side

CONVERSIONS

Time

365 days = 1 year
12 months = 1 year
52 weeks = 1 year
7 days = 1 week
24 hours = 1 day
60 minutes = 1 hour
60 seconds = 1 minute

Length and Area

5,280 feet = 1 mile
1,760 yards = 1 mile
3 feet = 1 yard
36 inches = 1 yard
12 inches = 1 foot
144 square inches = 1 square foot
4,840 square yards = 1 acre
1,000 meters = 1 kilometer
100 centimeters = 1 meter
1,000 millimeters = 1 meter
10 millimeters = 1 centimeter

Weight

2,000 pounds = 1 ton
16 ounces = 1 pound
1,000 grams = 1 kilogram
1,000 milligrams = 1 gram

Volume

4 quarts = 1 gallon
2 pints = 1 quart
4 cups = 1 quart
2 cups = 1 pint
32 ounces = 1 quart
16 ounces = 1 pint
8 ounces = 1 cup
1,000 milliliters = 1 liter

Metric to Customary

1 kilometer = .62 mile
1 meter = 39.37 inches
1 centimeter = .39 inch
1 liter = 1.057 quarts
1 kilogram = 2.2 pounds
1 gram = .035 ounce

Glossary

A

approximation An estimate that is close to a given number, but not exact. A bag of fruit marked $2.89 will cost approximately $3.00.

arithmetic operations Addition, subtraction, multiplication, and division.

C

combination word problem A word problem that needs 2 or more steps to be solved.

conversion Changing from one type of measurement to another. 12 inches = 1 foot.

D

denominator The bottom number of a fraction. In the fraction $\frac{1}{2}$, 2 is the denominator.

diagram A picture or visual aid that helps you understand a word problem.

E

estimation An approximate amount used to determine the accuracy of the arithmetic, or used to give an idea of the answer before doing the arithmetic. 17 + 52 can be rounded to 20 + 50 so the estimation is 70.

F

formula An equation that states a rule or factual information that can be used to solve a certain type of problem. $A = lw$ is the formula for finding the area of a rectangle.

G

given information All the numbers and labels that are in the word problem.

K

key word A clue that can help you decide which arithmetic operation to use. In the problem, find the difference between 15 and 6, the key word *difference* helps you decide to subtract.

L

label The noun (word or symbol) that a number refers to. If you are adding 6 apples and 5 apples, the total amount will be 11 and the label will be apples.

M

math intuition A general understanding of numbers, math operations, and a feel for what a solution should be.

mental math Arithmetic operations that can be done in your head, without the use of pencil and paper or calculator.

N

necessary information The numbers and labels in a word problem that are needed to find a solution.

number sentence A restatement of a word problem as an equation using numbers and labels. In the problem, find the total amount spent for lunch if Jody spent $4.29 and Doretha spent $6.45, the number sentence would be $4.29 + $6.45 = total $ for lunch.

numerator The top number of a fraction. In the fraction $\frac{3}{4}$, 3 is the numerator.

O

order of operations Rules that govern the sequence when you must use more than one arithmetic operation. Order of operations: do arithmetic in parentheses first; do multiplication and division before addition and subtraction; solve from left to right.

P

part A piece that is being compared to a whole.

percent A part of a hundred.

percent circle A memory aid used to help solve percent problems.

proportion A math equation that states that two ratios are equal. $\frac{2}{4} = \frac{1}{2}$ is a proportion.

Q

question The part of a word problem that tells you what to look for.

R

ratio A comparison of the relative size of two groups. The ratio for 3 teachers working with 15 students is 3 teachers/15 students.

restating a problem Saying a problem in your own words to help understand what the problem is asking.

round Estimate to a particular place value. The number 779 rounded to the nearest hundred is 800; rounded to the nearest ten is 780.

S

solution A number and label that will correctly answer the question.

solve To find the solution.

substitution Temporarily replacing a difficult-to-understand number with a small whole number that is easier to picture.

W

word problem A sentence or group of sentences that tells a story, contains numbers, and asks the reader to find another number.

whole A complete amount; in percent, the base for comparison.

Index